Economy–Environment– Development–Knowledge

As we approach the next millennium, we find ourselves in times of radical social change. Theoretical explanations of the economy, the environment and the development process furnish a plethora of policy prescriptions for such issues as employment creation, environmental degradation and social progress. Different policy strategies are compatible with distinct political programmes.

Economy–Environment–Development–Knowledge addresses alternative perspectives on these fundamental aspects of human existence. Economists, environmentalists and development theorists are unable to agree on how to address problems, or even on what the problem *is*. To understand and compare alternative understandings of economic, environmental and development issues, we need to be aware of *why* theorists conceptualize the process of social experience so differently.

Part I of *Economy–Environment–Development–Knowledge* addresses the subjective preference, cost-of-production and abstract labour theories of value in economics; Part II explains egocentrism, ecocentrism and sociocentrism as competing theoretical perspectives in environmental theory; Part III highlights modernization theory, structuralist theory and class struggle as ways to account for the process of development; and Part IV examines the generation of knowledge through positivism, paradigms and praxis, legitimating competing perspectives in economics, environmentalism and development. The book concludes by considering why different people find alternative explanations more or less plausible, and what the political implications of alternative analyses are.

By addressing the disagreements between theorists, *Economy–Environment–Development–Knowledge* provides a unique basis to contrast and compare the range of theories of, and policies for, economic prosperity, environmental sustainability, and social progress.

Ken Cole is a senior lecturer in the School of Development Studies at the University of East Anglia, Norwich.

Economy–Environment–Development–Knowledge

Ken Cole

London and New York

First published 1999
by Routledge
11 New Fetter Lane, London EC4P 4EE

Simultaneously published in the USA and Canada
by Routledge
29 West 35th Street, New York, NY 10001

Routledge is an imprint of the Taylor & Francis Group

Typeset in Galliard by
The Florence Group, Stoodleigh, Devon
Printed and bound in Great Britain by
Biddles Ltd, Guildford and Kings Lynn

British Library Cataloguing in Publication Data
A catalogue record for this book is available from the British Library

Library of Congress Cataloging in Publication Data
Cole, Ken.
  Economy–environment–development–knowledge / Ken Cole.
  p.   cm.
  Includes bibliographical references and index.
  1. Environmental economics. 2. Economic development. I. Title.
HD75.6.C6413   1999
333.7–dc21   99–28165

ISBN 0–415–16258–0 (hbk)
ISBN 0–415–16259–9 (pbk)

For Jenny and Alex

Contents

Acknowledgements

This book reflects my intellectual development over the past twenty years or so, and I owe a debt to all those who have played a part in that process. In particular I have profited from the privilege of working with so many students in my role as 'teacher', though often their intuitive, insightful comments acted as a stimulus to further thought and consideration, raising the question as to who was teaching whom.

I have also been fortunate to work with and teach ANC activists from South Africa before the fall of apartheid in 1994; Cuban economists and social planners and thinkers in Havana after the collapse of the Soviet Union in 1991 and the tightening of the United States inspired economic embargo between 1991 and 1994; Canadian rehabilitation professionals working with the disabled in Bosnia after the war and the experience of ethnic cleansing of 1991–94; and this wealth of experience has heightened my sense of intellectual responsibility. Knowledge is *power*. It is a weapon to be used with care. This book is intended to make explicit the moral, ideological, and ultimately political implications of alternative approaches to the understanding of social experience in general, and in the realms of the economy, the environment, and the development process in particular. In as far as this has been achieved, in no small measure, it is a consequence of the sensibilities and intuitions that have been honed by comparing and contrasting the musings of a privileged academic from England with the privations and dangers suffered, and the commitment and solidarity shown, by my South African, Cuban, Canadian and Bosnian friends.

I have also benefited from the interest of, conversations with, and incisive comments from Phil O'Keefe of the University of Northumbria at Newcastle, George Lambie from Leicester De Montfort University, Ian Yaxley working at Queen Margaret College, Edinburgh, and Gilberto Valdés and Humberto Miranda in the Instituto de Filosofia in Havana.

And finally, Jenny, my wife, and Alex, my son, have shown patience beyond the call of duty in living with an academic preoccupied with ideas.

Prologue

What *can* we know about the world?

Paradox, dominance, apathy and democracy

This book has grown out of a concern to explain human behaviour and social change, and to be able to understand what 'progress' might be and how it might be realized. But here is the paradox. One person's progress is another's regression. Perceptions of reality are distinct.

> The phenomena we describe and purport to explain appear to be constructed according to hypotheses derived from our own fallible senses, cultural traditions, social expectations and limited technological powers.
>
> (Rose 1997: 69)

Differences in individuals' consciousness of their social existence have a moral complexion. People's differing attitudes, intuitions, and behaviour come to be assessed as *good* or *bad*: *progressive* or *reactionary*. And it is precisely such a moral clime that stimulates activity to change society for the better, and yet in that moral imperative conflict over objectives seems to be inherent. Improvement for some is recessive for others, and the conflict over 'progress' is engaged in the moral fervour of people feeling that the order of things is not as it *should* be.

For people to make up their minds over what *should* be, there has to be an appreciation of what has been, and a conception of human potentials for an awareness of what is possible. People's choices *today*, over what to do *tomorrow*, are based on an understanding of *yesterday*. But the more we understand our past experience – an experience that we have shared with, and in which we have related to, conflicted over and compromised with, other people – the more complicated the present appears to become. Perceptions of reality are diffuse and complex. And to comprehend complexity, to understand our past to organize our future, experience has to be simplified: the decisive features of experience abstracted out of the confusing maelstrom of social existence; a differentiation consequent on the conceptualization of the possibilities of human activity; a belief in human potential. An understanding of human nature, and human motivation.

So, intrinsic to an understanding of behaviour and change is an awareness of the disparity in perception of experience. And the disparity in perception implies different intellectual parameters as to human potentials, parameters through which experience is perceived, and hence understood. This raises the question: How can complex behaviour be explained by simple postulates? And how are these postulates legitimated by the philosophical propositions of scientific inquiry, which invest analyses with credibility? But the pursuit is more than this. Theorizing experience, change and development, might rest upon the simplification of complex patterns of social experience between people, but every day we experience the world as a simple place. We have to survive.

> So, in order to comprehend our world and humanity's place within it, we must do more than just explain higher-level complexities in terms of lower-level simplicities. We must explain why, on every level of existence, we can deal with the world as *if it were simple*.
>
> (Cohen and Stewart 1994: 1, emphasis added)

How do we begin to understand multiple, accurate, perceptions of the same reality based on people's understanding of their differential experience? Perceptions that to the perceiver are reality, and which inform individuals as to the appropriate behaviour concordant with their ambitions to progress, to develop? In sum, how can we define the shared 'real' world, and what meaning we can give to 'development'?

The inability to take cognizance of the rich variety of people's experience and realities and the resultant plethora of ambitions and objectives for the future, and *their* accounts and descriptions of the past, leads to: *dominance* – the imposition of the views of the powerful on the weak; *apathy* – a fatalistic acceptance of the status quo and concomitant culturally sanctioned images of reality, as being either too complex to understand or too entrenched to change; or *democracy* – the political organization of society to reflect contrasting points of view. Apparent inconsistency politically resolved into symmetry.

The implication for the political organization of society of alternative conceptualizations of social experience and social change, and different understandings of democracy, is an underlying theme throughout the whole book, being explicitly addressed in the final chapter.

However, while being conscious of different conceptions of reality, the argument of this book is not positing a 'common-sense' relativism: that there is nothing to choose between alternative, theoretical simplifications of experience – alternative 'real worlds'. Rather, we do all share the *same* world, although it appears in different guises to different people, and hence social reality has to be theoretically constructed from the various perceptions taken from different social vantage points. The real world cannot be observed, but is discovered *through* observation. The scope of this book is ambitious, and necessarily the arguments are complex: as is social existence! To aid intelligibility, I have summarized the argument at various points in the book – the first pages of Chapters 1 and 18, and the first page of each of the four sections of the book – in the form of *flow charts* (what Tony Buzan (1974, 1995) calls *mind maps*) in an attempt to make the complex logical development of the argument explicit.

I hope you find what follows stimulating and interesting.

The argument in outline

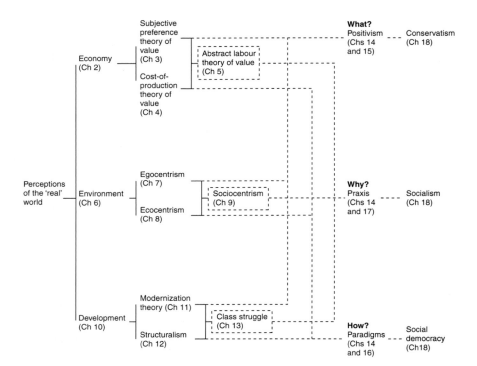

1 The scientific parameters of social existence

Perceptions of reality

The prologue ended by postulating that the real world cannot be observed, but is discovered *through* observation. What does this mean?

People's perceptions of reality reflect their experience, an experience that they have to understand if they are to be able to better organize their activity if they are to improve that experience in order to progress. But individuals do not change society, social change is a *social process*. And purposeful social activity to affect change has to address and be relevant to a whole spectrum of personal experiences and perceptions of the 'real' world. So, in theorizing 'development' we have to take account of people's differing intuitions and conceptions of what the world *could* become, to attenuate the problems and frustrations of the present. We have to ask the question: What must the 'real world' be like if people's experience and perception of that world is so different?

As intellectuals, as theorists, we have to scrutinize the various representations of reality. After ensuring that the various understandings of development are an accurate articulation of someone's experience, that the understanding is logically coherent and empirically accurate, we then have to take that perception seriously, as valid. In as far as we are trying to explain human behaviour, and humans act to change the world according to what *they* think that world is, and what *they* think human potentials are, then we have to incorporate people's *own* understanding of their experience and *their* beliefs as to humans' potentials, into the theorization of the real world and the potentials for change.

As we shall see below, the various experiences of social life and perceptions of reality fall into distinct perspectives, visions of human life that have a moral, ideological and political dimension. And it is this implicit agenda that it is the project of this book to understand.

This chapter gives a rapid overview of the development of the argument in the subsequent seventeen chapters. In places the argument is complex, because the reader is asked to understand and evaluate alternative explanations of essentially the *same* experience. Inevitably this requires that credulity is suspended to reveal consistency; that our own perception of reality is abrogated so that we may know others' experience better, thereby expanding self-awareness. Such extended self-consciousness is a prerequisite for a meaningful understanding of our lives, and the choices that may be made over how to better organize the future: development.

Throughout the book the understanding of change is explicit, but implications of this theme as to how to prosecute rational change are only finally made explicit in Chapter 18, where the book concludes on the politics of knowledge.

Economy

Chapter 2 begins by considering what economists think the economy *is*: economists operate within different theoretical perspectives, perspectives that define the nature of economic relations differently, and work within distinct intellectual parameters for the definition of economic reality.

> '[M]ost controversies on economic theory involve a clash of assumptions about the structure of economic *reality*.'
> (Bell and Kristol 1981: viii, emphasis added)

People live in societies characterized by a technical division of labour; they specialize in producing particular commodities. They do not produce what they consume, nor consume what they produce: hence there has to be exchange between producers and consumers. Economists try to understand the terms of this relationship of exchange: the *value* of commodities. But different perspectives on economic theory define value differently. There are different definitions of value, different theorizations of how value comes to be, and different conceptualizations of the dynamic of relations of exchange. Do the terms of exchange reflect the preferences of consumers, the costs borne by producers, or the power of citizens to influence the value and price of commodities consumed and produced?

Is the consumer, the producer, or the citizen, the *dynamic* of economic activity? Within these distinct perspectives, the same words, such as 'price', 'value', 'supply', 'demand', mean different things. Although the word is the same, the concept is not. Hence economists working within different perspectives are essentially operating with different languages, leading to distorted communication between theorists. An example of such confusion is given by addressing different interpretations of 'that most basic of economic concepts': the theory of price formation by the forces of supply and demand.

Chapter 3 addresses analyses of the economy where the consumer is assumed to be the economic dynamic. Individuals are believed to be biologically endowed with a unique set of tastes and talents, which define individuals' subjective preferences for maximising utility (being happy) – which defines 'value'. This is the *subjective preference theory* of value. As people are independent individuals, their preferences for utility can only be expressed if they exercise free choice, in unrestricted markets.

Austrian economic theorists try to understand how economies actually function, with markets conceived of as information mechanisms by which individuals learn of each others' tastes, and talents. And since individuals' preferences are unique, and it is unknowable what people will decide to do, it is impossible to define an equilibrium state where everyone's utility is maximized.

Neoclassical theorists, on the other hand, believe that such a state *can* be defined: indeed it is the function of governments to create an environment of *perfect competition* through which utility will be maximized. Economic policy then is intended to make the real world more and more like the perfectly competitive *utopia*. It is impossible to determine what prices and earnings *should* be, since tastes and talents are unique to the individual, and the only conclusion that can be definitively drawn is that as long as free markets obtain, then people have only themselves to congratulate or blame for their good fortune or their predicament. However, the theory of the *second best*, and the *capital controversy* (addressed in Chapter 4), highlight potential logical contradictions in the proposition that free markets unambiguously lead to the maximization of individuals' utility. And the belief that progress is attained through free exchange remains just that: a *belief*.

The role of the state in prosecuting economic policy is addressed, and the issues of inflation, employment, and public goods and environmental economics, are specifically discussed. In Chapter 4 we reconsider the determination of value, and the rate of exchange between consumers and producers, this time assuming that the *producer*, and not the consumer, is the economic dynamic. We start with Alfred Marshall's proposition that the cost of production determines value, and not consumer satisfaction. Production is characterized by a technical division of labour, rendering individuals dependent on each other. People have to organize themselves to cooperate with each other; individuals are not 'free to choose' and compete in markets to maximize personal utility.

The concern is not now with the microeconomy of consumer choice, but with the macroeconomy, and the management of cooperative economic activity. For the *cost-of-production theory of value* there are essentially two policy imperatives: the technical question of the efficient utilization of productive resources, and the distributional question of 'fair' rewards. For the subjective preference theory of value, in competitive economies, individuals' incomes reflect their individual talents; for cost-of-production theory, on the other hand, the concern is to redistribute extreme wealth to relieve extreme poverty, creating a social environment conducive to economic cooperation. In this regard, economists derive policy prescriptions from considering the 'actual' economy, not by deriving economic objectives from the utopian model of perfect competition.

In an attempt to conceptualize price formation more accurately, models of *imperfect competition* were developed, in which market prices reflect producers' decisions rather than consumers' preferences. However, it was not until the work of John Maynard Keynes that cost-of-production theorists had a model of how whole economies reflect producers' exigencies. And Keynes' rationalization of government – economic intervention – had distinct political implications for state activity, although these economic obligations were moderated by the Keynesian reinterpretation of Keynes' problematic according to the intellectual (and political) parameters of the subjective preference theory of value.

However, the logical contradictions of trying to argue for, and theoretically justify, free markets *and* economic management finally came to a head in debates around the *capital controversy*, which fundamentally called into question the whole theoretical edifice that separates economic theory and economic problems into micro and macro issues. However, it is still extremely common for economists, and economic courses and textbooks, to conceptualize economic issues in this anachronistic, illogical, terminology. And governments still try to 'manage', in the social interest, 'free' markets that reflect individuals' preferences. But the juxtaposition of the concepts 'manage' and 'free', and 'social' and 'individual', is the portent of theoretical contradiction.

There has been a theoretical resurgence of the cost-of-production theory of value in the light of the policy emphases of the subjective preference theory of value in the 1970s and 1980s, which, focusing on prices and inflation, and emphasizing the competitive dynamic of individuals maximizing utility, presided over an increase in economic inequality and the underutilization of economic resources and rising unemployment. The resultant political opposition calling for economic intervention is ideologically legitimated by the arguments of the cost-of-production theory of value, the theoretical rationale being based upon the assumption that producers, and not consumers, are the economic dynamic.

The argument of Chapter 5 is that it is not merely the preferences of independent consumers, nor the technical exigencies of dependent producers, that is the economic dynamic, but it is the relations of power between interdependent 'citizens' that explains economic activity. With regard to relations of exchange, the level of people's control over the social means of production limits the extent to which they can choose to consume, or organize the management of production. People's relationship to the means of production defines their 'class' position in society, a position that not only confers power in the organization of economic activity, but puts them in a pivotal position to influence society in general, as cultures emerge that fundamentally justify the social status quo based on class power.

Class power in the capitalist mode of production is essentially effected through *commodity exchange*, and the concept of the 'commodity' is addressed, highlighting the process of *exploitation*, by which the commodity 'labour power' is valued by the market at *less* than the product of wage labour. And this surplus value is distributed as the incomes profit, rent and interest via market exchange. Value itself does not exist until the *social* process of exchange: a process by which the product of individuals' labour is abstractly valued by society – the *abstract labour theory of value*.

And because value is a consequence of social relations of exchange, which *follow* production, there can be economic crises, where producers do not realize the expected value from market transactions. While competitive market relations makes crises possible, the class-based, inherently exploitative nature of production make crises inevitable. The competitive dynamic of the world, the capitalist mode of production, placing a premium on economic profitability and efficiency, inexorably leads to a trend for profits to decline: a trend that may be offset by action to increase the rate of exploitation of wage labour. As and when profits actually decline, and enterprises cannot remain economically viable, the effect on people will reflect the class power of labour and the material conditions of production in different parts of the world.

Finally, Chapter 5 addresses the process of crisis in the world economy in 1997 and 1998.

Environment

Over the past twenty years or so ecology has become ever more pervasive as an issue in world politics. Chapter 6 begins by charting the course of the emergence of the environment as a political issue.

But just as with economic policies, which differ as a consequence of distinct concepts of value, so with environmental prescriptions. Environmental and ecological theorists differ over the conception of the environment. Is the environment essentially a resource to 'consume'; or is it a 'productive' resource; or do the environment and human society co-evolve, and the important questions are over the justice of the way the environment is defined and utilized for human needs?

Different intellectual parameters ask different questions about the environment. What is it? How is ecological balance maintained? Why are there environmental problems? Consequently there are distinct conceptions of sustainability. And different approaches to science define 'relevant knowledge' differently. Although knowledge is not explicitly addressed until Part IV of this book, Chapters 7–9 address the limits of the intellectual parameters of environmental theory.

Chapter 7 focuses on the environment as a source of utility for consumers: *egocentrism. The* environmental problem is one of *conservation*. If the environment is defined in terms of consumption, problems reflect the numbers of consumers: *population growth.* Neo-Malthusian theorists are very pessimistic as to the future of human society, as the population grows inexorably, resources are depleted, and the planet cannot absorb spiralling pollution. However, not all egocentric theorists share the apocalyptic vision of the neo-Malthusians. It is argued that environmental problems will be experienced as shortages, and as long as there are 'free' markets functioning with regard to the consumption of the environment, competitive markets will allocate resources appropriately.

The environmental problem is the 'correct' valuation of the environment, and letting the forces of self-interest manage the environment through the invisible hand of the market. And in this regard there has been a debate over the allocation of property rights to ensure that utility is maximized in the exploitation of the environment.

There is a continuing debate over how far, and the method by which, the state can intervene to control the exploitation of the environment. Again the emphasis is on correctly valuing environmental assets through the technique of *cost-benefit analysis*. The discussion centres on how the environment can be valued when it isn't marketed as a commodity, and the means by which the future effects on consumers' utility of today's activity are valued. In this debate the issues raised in Part I of this book, where the problems of the second best and the capital controversy were highlighted, are conspicuous by their absence, seriously questioning the intellectual coherence of this approach to environmental economics.

Essentially the concern is to *commoditize* the environment, and in this context an optimum level of pollution can be defined, at which point individuals' utility is maximized. And within these optimum limits, permits to pollute can be traded as assets, which for egocentric, environmental theorists is a 'great leap forward in environmental thinking'.

In Chapter 8 the concern is not with the environment as a source of consumers' utility; indeed, for *ecocentric* theorists, who understand the environment as an ecological system that is delicately balanced, it is precisely treating the ecology as a source of pleasure for individuals that has created present-day ecological problems such as global warming, pollution and the destruction of biodiversity.

The ecology is essentially a productive resource, and has to be managed according to the technical exigencies of the production process. Human existence is ecologically conditional, and individuals have to learn to live within the natural and technical limits of the environment.

The *policy pragmatists* do not share the faith of the *egocentrics* that humans will spontaneously respond to market signals, even if prices can be manipulated to create incentives to conserve the environment. Social institutions specifically intended to maintain the integrity of the environment have to be designed to address particular problems. However, the *Gaia* theorists, while agreeing that social life has to be managed to stay within ecological limits, believe that our ecological knowledge is far too primitive, and that the management decisions and policy prescriptions are likely to make the situation worse rather than alleviate environmental problems. And we have to learn to live within natural limits, an awareness created through appropriate education.

The term 'Gaia' comes from the pre-classical Greeks, who worshipped the Earth goddess Gaia. And some Gaia theorists verge on a religious commitment to 'save the earth'. Education is not enough, people have to 'see the light'; they have to possess religious conviction and vision. Some ecocentric theorists are more explicitly stirred by religious passions, and are more *theo*centric than *eco*centric. The theorist has to become an evangelist.

Such a religious awakening characterises *deep ecology*. Everything that is a part of the ecology is believed to have a right to exist. The maintenance of biological diversity is an end in itself. The *social ecologists* combine the insights of the deep ecologists with the scepticism of the Gaia theorists. It is our ecological mismanagement that has created our present predicament, and having decided how we *should* relate to the ecology, we have to decentralize power into a network of local assemblies so that we can manage our lives on a scale consistent with ecological limits.

Political ecology goes a step further. We have to create institutions to resolve conflict between people's differing attitudes towards life, which are based on their distinct experiences, giving rise to different theoretical discourses. Poverty is intimately linked to ecological problems but, ultimately, because there is a common interest with regard to the use of the environment, we need an intellectual space within which we can theoretically resolve contradictions in our understanding of the ecology.

Pursuing the issue of contradictions within ecological theory, the *liberation ecologists* see the contradictions as much more political than logical. Poverty might lead to environmental degradation, but poverty itself is the flip side of the coin of wealth.

Chapter 9 really takes this insight as its theoretical springboard. But the evolution of the environment is seen as much as a social process as a natural process. People are not only interdependent with each other, but also with the natural environment: there is a *co-evolutionary* process of change. The ecology is socially defined, and vice versa. Ecological crises are crucially linked to the 'commodification' of the environment, a consequence of production within the capitalist mode of production. The environmental, social and economic consequences of social organization according to the exigencies of commodity exchange make the economic exploitation of the environment an inevitability.

The environment sets the natural parameters to social existence and change; and changes in the basis of social existence impact upon the environment. The relationship is dialectical. And because people are creative, self-conscious beings, the relationship is indeterminate, but potentially *can* be understood and consciously transformed with the development of class-consciousness.

The theorist becomes the activist. And the political context of class-based social change is the constraint on people's fulfilment of their potentials created by commodity exchange. Ecological problems are socially created. Nature has no 'will'; no choices are made by nature between ecological alternatives. It is self-conscious humans who choose – choices based on an understanding of past experience. But many theorists see the contradictions of social existence as being sited in natural forces beyond human control. However, for Karl Marx the contradictions frustrating people from fulfilling their social potentials were not beyond human influence. And action to remove the constraints on people's potentials is consequent upon individuals' being conscious of their class interest. The ecology is not a system, but an aspect of a changing *process* of social existence, through which people fulfil their changing potentials in the context of unequal power: development through conflict.

While environmental degradation might be a consequence of industrial development or poverty, it is not *caused* by industrial production or inequality. And changing society crucially reflects people's awareness of their class interests, which in turn reflect the ideology that colours their perceptions of the world. But how do such perceptions change? Scientific theory and the activity of scientists are also social activities moulded by social experience, riven by conflict: conflicts that are rationalized as alternative principles for social organization. People with shared interests mobilize and organize themselves to effect social change.

Development

Part III (Chapters 10–13) turns our attention to the development process. As for economists and environmentalists working within different perspectives, who cannot agree on the definition of 'value' or the nature of the environment, development theorists work within distinct theoretical, intellectual parameters as to the nature of 'progress'.

Development implies social change consequent upon people changing their behaviour. And development theory must be based on a conception of human motivation as to why individuals should act in concert to effect a change in society. Is it that individuals have to be 'free' to fulfil their potentials; or does change have to be managed in the social interest; or are there contradictory interests within society that are politically prosecuted to the advantage of the powerful?

Does development involve a process of the modernization of society to allow individuals to be free *and* the modernization of people's minds rendering them individually responsible for their interests? Or does progress require more effective management of social and economic activity in the general interest? In this latter regard there has been considerable theoretical work in defining and measuring development to provide the basis for effective planning and management. However, that people have differing interests *and* have to cooperate in social existence can be conceptualized through activity being orientated towards the maintenance of class interests. And purposeful change is then a consequence of people being class conscious, and being aware that development is synonymous with a redistribution of class power. The development theorist is at one and the same time a political activist; and 'progress' reflects people's participation in the organization and control of social and economic life.

Chapter 11 addresses in more detail development as a process of modernization. Because individuals' potentials are innate, the concern is for people to be aware of what they can do if they are allowed to be free, and in creating a social environment free of restrictions: the modernization of the mind and of society. At the heart of such a social theory of development lies the belief that universally, as people become aware of their potentials, there will be a struggle to be free; and all societies go though essentially the same development process. Progress is the transformation of traditional society into a modern state through various stages.

According to *dualist theorists,* such a transition from a traditional to a modern society might be effected by the introduction of market rationality through a programme of *import substitution industrialization,* though in modern times the mantle of modernization theory has passed to the International Monetary Fund and the World Bank under the guise of *structural adjustment.* Economic growth is understood to be the harbinger of social progress, and growth is best achieved through competitive capitalism and market forces. But the disappointing effects of structural adjustment programmes, which have not unambiguously produced something that might be called 'progress', have prompted *new institutional economics* theorists to try to identify institutional attempts to give security in an uncertain world as unintentionally frustrating enterprising individuals from fulfilling their potentials. And in this regard attention has shifted to an environment of good governance, which guarantees individual freedom in general and free exchange in particular, as a prerequisite for development and progress.

To generalize good governance to the world economy there is pressure to guarantee the rights of private, international investment to freely enter any branch of economic activity, under a protocol called the Multilateral Agreement on Investment.

Chapter 12 shifts the emphasis of responsibility from the individual to society. Development is a process of fulfilling the technical potentials of production, which because of the technical division of labour requires improved cooperation between individuals through appropriate social institutions. The structural, institutional basis of managed, social and economic life has to change. And clearly for there to be social cooperation to effect development, in a context of individuals depending on each other for their well-being, there has to be a culture of fairness. And theory must be based on a detailed analysis of the actual ways in which people cooperate to survive socially.

In this regard, *populist* theorists highlight the problems of particular groups of disadvantaged people who share a distributional interest. Often these problems are a consequence of processes of industrialization, which marginalize the powerless. And some structuralist theorists argue that the economic and productive, and the political dimensions of social life should be appropriate to, and decentralized to create, a world that is understood and managed by people themselves. For other structuralists poverty can only be resolved by economic growth, and society has to change to embrace a strategy that will combine economic advance with a fair distribution of benefits. And such an option is consequent on political will and rational argument about the technical prerequisites of social existence.

However, not all structuralist theorists are so sanguine about the potential for rational politics to overcome the powerlessness effected by technical underdevelopment; nor of the will of political elites to organize meaningful social and economic change. For such theorists economic dependency is too much of a constraint to be amenable to rational politics.

For other thinkers the failure of such a technical-based strategy to result in or explain development, or the lack of it, over the last twenty years or so, throws into question current theories. Existing conceptions and theoretical explanations, the extant development paradigm, have to be amended to account for the new global reality of economic activity. For these theorists there is a theoretical impasse: the world has changed, but our understandings have not.

Chapter 13 addresses the development process from the perspective of individuals struggling to fulfil their potentials, in contexts where social, class power is contingent upon control of the means of production. People's activity no longer is a reflection of their genetic inheritance, nor is behaviour simply explained by individuals' having to adapt to the needs of society. Rather, people's biological characteristics only become potentials *through* social experience; but more than this, people's social potentials *change* with social experience. People are uniquely creative, and development reflects the social opportunity to realize individuals' potentials – opportunities that are constrained by class power.

Progress is consequent on a redistribution of power, and normally such a realignment is the product of political conflict, though often such struggles are not conceived of in 'class' terms. A purposeful development strategy requires people to be 'class conscious' to interpret issues of resource allocation, education provision, employment, etc. as issues of class inequality. And such everyday frustrations to people's fulfilling their potentials are not obviously class issues. The concept of 'class' does not describe lifestyles, but explains interests; and class-consciousness is not something that people *possess*, it is something that people, socially, *do*. It is a question of people understanding that they share fundamental interests with other people, and that they are mobilized and organized to act in concert to realize shared objectives against a ruling class with contradictory interests.

The ideal, where everybody could potentially fulfil their unique social potentials, is *communism*. People change through their experience, and hence the communal operation of production would require a 'different kind of human material'. And yet communism is *emergent*: it is a consequence of the contradictions of capitalist society, exhibiting the moral, cultural and economic traits synergetic in the capitalist mode of production. Socialist development, based on the increasing participation of creative individuals, is the process

by which people's nature changes as they better fulfil their changing potentials. The development theorist becomes the political activist, as through
praxis people become increasingly class conscious. And through education
and increased awareness, people liberate themselves as they are able to conceptualize and articulate their intuitions and insights, gaining the confidence and
ability to organize themselves to challenge the status quo. In this context
the case of socialist development in Cuba is addressed.

Since the revolutionary victory of 1959 Cuban development has been driven
by the need to industrialize, and the imperative for people to fulfil their
potentials. And amongst developing countries Cuba has an enviable record
in health services, education provision and poverty relief, in spite of the unre-
lenting US-sponsored economic blockade, and the collapse of Cuba's one-time
ally, the Soviet Union, which at a stroke denied Cuba export markets and
sources of imports – referred to as the 'double blockade' in Cuba. However,
Cuba was never a Soviet satellite, the Soviet bloc becoming a vehicle to
escape sugar dependency and provide succour in the face of US aggression.
Various phases of development policy in Cuba have reflected the opportunities for economic development, and the need for power to be decentralized,
based upon economic equality and political participation. In recent times the
political and economic evolution of Cuba has been constrained by the effects
of the double blockade.

But still the *process* of 'socialism' in Cuba continues.

Knowledge

In Parts I, II and III of this book, addressing alternative understandings of
the economy, the environment, and development, the different perspectives
are each, potentially, able to offer coherent explanations of the 'same'
experience – each is *realist*; the alternative analyses are logically coherent,
proceeding from assumptions to conclusions – each is *rationalist*; policy
prescriptions are devised to ameliorate problems – each is *activist* (on the
realism, rationalism and activism of competing theoretical perspectives, see
Cole *et al.* 1991, Chapter 1). The issue then is not one of discovering the
neutral, 'scientific' analysis, but one of understanding the differences between
competing perspectives and the significance of these emphases for the explanation of social activity, and hence for people's lives.

Different theories of what constitutes 'knowledge' underlie the various
analyses, each of which can be considered to be scientific, the various theorizations being legitimated by distinct concepts of science. We have then to
turn our attention to defining 'knowledge'. The first question then is: What
constitutes knowledge?

Chapter 14 considers the nature of the human mind: Why do humans
strive to understand their experience, and why in this endeavour should there
be distinct intellectual perspectives? The same experience can be interpreted
differently by different people: effectively, people's 'real' worlds are distinct.

And these alternative conceptions of the world are legitimated as scientific by reference to theories of knowledge, and therefore justified as valid by different *ideologies* of science. There is then no universal agreement as to what constitutes legitimate knowledge, no shared standard of objectivity between competing perspectives. Consequently there is distorted communication *between* scientists, who effectively talk different theoretical languages drawing different inferences from shared experience. And yet, we all share the *same* world. How do we begin to define what this 'real world' is?

The question is to differentiate between one thinker's subjective interpretation and another's objective reality.

Humans are unique in the degree to which they are *self*-conscious: people are aware of themselves, aware of being alive. Being aware of themselves, people can reflect on what they have done, and consider alternative scenarios of what is to come; people can imagine what they will do tomorrow. Hence choices can be made intended to make tomorrow better than today; humans purposively strive to 'progress', to *develop*. However, individuals' choices will reflect their understanding of what their potentials might be, and what have been the significant influences on their life hitherto – what has structured and constrained their experience. And it is people's consciousness as to what they in particular, and humans in general, are capable of, and the determinants of the dynamic of their lives, that structure these choices. Fundamentally, humans are *curious*.

The only evidence upon which such a consciousness can be based is life – their own plus possibly an awareness of others' lived experience – and such experience is *social*. The need to understand social life is a uniquely human attribute, a consequence of our being self-conscious.

Of course, the ability to self-reflect on our lives went hand in hand with the evolution of the human mind, which has given humans spectacular survival advantages over other species. With the emergence of the mind came symbolic representation: *language*. Humans developed the facility to learn from each other. Progress and development do not have to wait for biological evolution. Change does not have to be a consequence of genetic adaptation to new ways of life, nor of genetic mutation. Progress and change can be very rapid. But such progress is contingent on understanding our experience: contingent on our 'consciousness'. People's understanding of what *has* been, and what *could* be, reflects their differential experience of what *is*. 'Tomorrow was born yesterday'.

The social context of individuals' experience, a world of abstract relationships, can only be theoretically appropriated based on beliefs about the nature of human motivation. Individuals' social behaviour can be understood to be a product of their genetic inheritance: biological determinism – nature. Or, perhaps people are 'socialized' to fit into society: social adaptation – nurture. Alternatively human nature could be a consequence of individuals' social experience offering opportunities to fulfil their biological potentials, and to strive for objectives that can only be reached in concert with other people. Human nature reflects individuals' social existence: the nature–nurture dialectic.

Are individuals independent of society; dependent on society; or interdependent within society? Beliefs as to the determinants of human nature imply distinct conceptions of knowledge, and a conception of scientific methodology and objectivity for the definition of 'reality'. Where the world is believed to be composed of independent individuals, the concern is to identify what happens, through a *reductionist* methodology. Where individuals are conceived of as being dependent on society, the emphasis shifts to identifying *how* events occur, through a *holistic* approach to science. If people are understood to be interdependent within society the problematic is one of understanding *why* rather than how events come to pass, through a *dialectical* approach to knowledge.

Chapter 15 addresses the identification of events, asking *What?* through a positivist conception of science. The concern is to establish the facts. But we can only know the facts by conceptualizing experience, which is achieved by reducing phenomena under investigation down to their individual components, so that we can identify significant events in our lives, and thereby manipulate what happens to improve our existence. Any effects extraneous to our objectives are ignored, and the analysis is only valid with regard to our limited objectives. Nothing can be said about a world of social relationships, or where people have distinct perceptions of that world reflecting their particular experience.

Individuals are necessarily judged to be independent of society with regard to their motivation. Behaviour is biologically determined. Theories are inaccurate descriptions of social life, being intended only to throw light on individuals' motivations. And 'objectivity' is defined as the *same* event being identified by different people, with theories judged to be scientifically accurate for as long as predicted events occur. Reality is contingent upon observation, and existing phenomena are correlated to one another, establishing a statistical relationship rather than a causal sequence of events.

Hence the scientific method is one of deduction from the facts, which are verified through empirical inquiry intended to test hypotheses. Phenomena have to be reduced down to their component parts to identify the particular event at issue, which is highlighted by statistical correlation. Such an approach to knowledge appeared at a particular point in history, when people increasingly were able to act according to their own volition, as the basis of social life and individuals' existence reflected market exchange. But the progress and the development of the human condition, which has been a consequence of individuals' acting ever more in concert to achieve shared objectives, has been increasingly inhibited by the positivist, reductionist, scientific methodology. Such limited intellectual parameters are unable to address relations *between* phenomena.

However, as an ideology, positivism has persisted in the social sciences, justifying social life according to the principles of *individual* exchange.

Chapter 16 considers the implications of *how* events are related for the definition of knowledge, and what should be the scientific methodology for investigating such interrelations. Phenomena cannot now be reduced to their individual components, but have to be situated in their relation to co-terminous phenomena. The approach has to be holistic, explaining the systematic interaction between events, which *only* have meaning in their juxtaposition with other events.

The scientist is not now concerned to deduce a statistical relationship from the verification of facts; rather, the scientific enterprise is to induce a causal relationship from an understanding of the systemic relation between phenomena. The approach to scientific knowledge is *paradigmatic* rather than positivist. Paradigms are essentially models of how systems work. And a paradigm is considered scientific for as long as the model explains a systemic relationship. However, when new problems emerge that cannot be explained by the extant paradigm, this heralds a period of 'revolutionary science'. At this time scientists are engaged in reconsidering and reformulating scientific knowledge, attempting to devise a new model, a paradigm, that can explain this anomaly.

Objectivity is not now a consequence of the accurate prediction of events, but an adequate explanation of systems. In recent times scientific, objective knowledge has seemingly proved inadequate for the explanation of social life, and the resultant period of revolutionary science has created considerable uncertainty as to how to organize and manage social existence. Such uncertainty has spawned a postmodernist approach to knowledge, which is unable to distinguish between individuals' perceptions of shared experience. These confusions reflect troubled times, when individuals are driven to defy social authority to defend particular interests according to their race, gender, ethnicity, sexuality, occupation, or any other of the particular conditions that structure their lives.

Chapter 17 looks in more detail at the implications for knowledge and science of the social context of existence, emphasizing *why* there should be theoretical anomalies, problems, that cannot be explained by paradigms that address *how* phenomena are interrelated. The concern is to understand *processes* rather than explain systems.

Consequently the emphasis is upon self-conscious people changing in response *to* their experience, rather than being explained *by* their experience. And now the parts (individuals) and the whole (society) co-determine each other: reductionism and holism are only partial approaches to understanding the complex, dialectical process of existence.

Individuals' actual perception of reality reflects their interpretation of social experience, and we can only understand and know what people's shared reality is if we theoretically juxtapose different interpretations to reveal the common bases of distinct experiences. The 'real world' cannot be observed, but is discovered *through* observation. As a consequence of social life people realize new potentials, and come to know themselves better as they understand others. As their individual consciousness changes, so their perception of life alters, and their 'actual' experience is different. Hence, the 'real world' that is discovered through the observations of different people changes, and reality is indeterminate. Scientific knowledge is now not about defining 'objective reality'; the concern is to identify potentials in an uncertain future.

There is a dialectic between objective experience and subjective consciousness, each determining the other. The social possibilities of human existence, the potential for progress, will reflect individuals' consciousness of *their* potentials: an awareness which changes with social experience as people realize new objectives. In a real sense people create themselves through their experience – an experience of contradictory interests reflecting the extent to which people control the process of social reproduction implying social and economic inequality, and frustration as individuals are socially constrained from fulfilling their potentials. And people are able to theoretically understand the uncertainties of their existence, which threaten to deny the disadvantaged the opportunity to fulfil their changing social potentials, through the 'science' of *praxis*. The social significance of individual experience is made explicit.

Intellectual panorama, ideological vision, and political view

Chapter 18 draws the sections on the economy, the environment, development and knowledge together, concluding not only that the mode of conceptualizing reality gives rise to distinct perspectives, but that these approaches are characterized by ideological biases. These ideological agendas define what *should* be the appropriate organization of society according to the implicit beliefs about human motivation that rationalize the analysis of social experience. *All* attempts to explain human activity must be based on a conception of human nature. Human existence is social existence, and it can never be definitively demonstrated if people's behaviour essentially reflects individuals' genetic inheritance (nature), their social experience (nurture), or if there is a dialectic between biology and society, with people discovering their potentials through social life (nature–nurture dialectic). Such a tripartite approach to defining human potentials underlies the various analyses produced on *every* social issue, and examples are cited from analyses of social behaviour across a range of intellectual disciplines.

In as far as individual behaviour is believed to be biologically determined, the people are essentially independent of society, and social life should be based on individual freedom. But where human life is understood to require cooperation within a technical division of labour, then individuals have to adapt to the needs of society; people now depend on others. Human nature reflects the process of socialization, and social life has to be managed in the general interest. Where the nature–nurture dialectic is understood to condition social activity, people are interdependent within society, and behaviour crucially reflects the social opportunities accorded to people to develop their potentials.

Identifying different understandings of human behaviour by reference to conceptions of human nature inevitably highlights the moral implications of theories. Knowledge itself has a moral dimension. And with regard to the theoretical definition of the 'real world', highlighted in Part IV on knowledge, the nature–nurture dialectical approach can be seen to be inclusive of the other two perspectives on human understanding.

The moral context of theory, and the theoretical definition of the 'real world', places the theorist in a pivotal role in society. Developing the analysis of Part IV on knowledge, should intellectuals be: *empiricists*, attempting to describe existence through a positivist analysis; *experts*, explaining the systemic relation between events through paradigms; or *activists*, developing *praxis* so that people can understand potentials of systemic relations to better fulfil their changing potentials?

Positivism implies a conservative bias in social organization; paradigm theory is consistent with social democracy; *praxis* implies a process of socialist development. Intellectuals have to be aware of the moral, ideological and political implications of scientific inquiry.

Part I
Economy

Economic thought

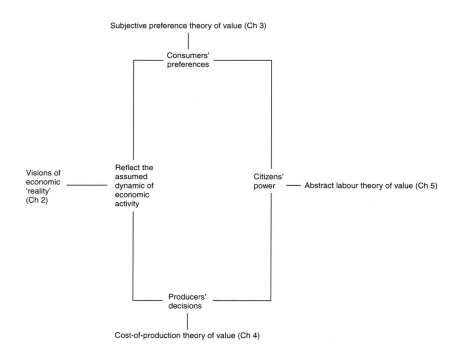

2   The economy

The economy and economists

Economists, those 'respectable Professors of the Dismal Science' (Thomas Carlyle 1850, quoted Sherden 1998: 55), find it very hard to agree on what the economy *is*.

> We talk about the same things, but we have not agreed what it is we are talking about.
>
> (Robbins 1984: 1)

Apart from the failure of economists to agree with each other, some seem to have problems deciding what they are doing themselves. For instance, Robert Heilbroner, an eminent economist and professor at the New School for Social Research in New York, and William Milberg, assistant professor at the New School, argue that '"powerful" economic theory is always erected on powerful sociopolitical visions ...': visions of 'political hopes and fears, social stereotypes, and *value judgements* ... that infuse all social thought'. And these visions structure our understanding, defining what 'we *believe* we know'. However, on the next page, they assert that economic theory is characterized by 'lawlike regularities of behaviour, investing it uniquely [within the 'social' intellectual disciplines] with the characteristics of a social "*science*"' (Heilbroner and Milberg 1995: 47, 4 and 5, emphasis added). How sociopolitical visions are reconciled to lawlike regularities is not explained.

As 'scientists', economists systematically observe economic behaviour, formulating general explanatory laws and hypotheses intended to predict the course of economic activity. And such predictions are used by policy makers, either in politics or in business, to chart an economic course intended to improve economic welfare, however defined.

However, not only are economists unable to offer sound guidance on day-to-day economic decisions, such as investment decisions and interest rates or the course of price changes, they are unable to predict *major* changes in economic activity.

Economic forecasters have routinely failed to foresee turning points in the economy: the coming of severe recessions, the start of recoveries, and periods of rapid increases or decreases in inflation . . . In fact, they have failed to predict the past four most severe recessions [in the United States] . . . economists' *vision* of the future is clearly clouded with situational *bias* . . . economic forecasters are heavily influenced by their particular economic *religion*.

(Sherden 1998: 55–6, emphasis added)

'. . . economic forecasts are the subject of open derision . . . their accuracy is appalling' (Ormerod, 1994: 3). Professor Paul Ormerod's comments are particularly relevant, as he was director of one of Britain's premier economic forecasting institutions, the Henley Centre for Forecasting. William Sherden has carried out a study of economic predictions in the United States, and found that: economists cannot predict the turning points of the economy; the inaccuracy gets worse the longer the lead time; economists' success is as good as 'guessing'; increased quantitative sophistication does not improve accuracy; there is no evidence that economic predictions have improved over the past three decades; and so on (see Sherden 1998, Chapter 3).

In this section of the book (Chapters 2–5, we shall consider the 'religious bias' of economists, and make explicit the beliefs that underlie the formulation of economic theory. Economists' visions define their perception of reality – a theme addressed more fully in Part IV. There will be no attempt to identify the 'best' theory, only to understand what the differences are, and what these distinct emphases imply for an understanding of the 'real world'.

Economists' differences

If all the economists in the world were laid end to end they would not reach a conclusion.

(George Bernard Shaw, attrib.)

People exist socially. Their social survival is characterized by being part of a technical division of labour: that is, it is more efficient and people can have higher standards of living if they specialize in particular aspects of production, which ideally reflect their particular abilities and talents. Consequently, people do not consume all they produce, nor produce all they consume. With low levels of technical efficiency the division of labour is within the household and the immediate locality. But with increased productivity and rising standards of living the division of labour goes beyond the household, homestead or village. Producers serve larger and larger markets, and exchange has to be regulated by other than social values and kinship obligations, and customs and traditions. Market forces are born, and products become commodities with a value and a price. There are *relations of exchange* between consumers and producers.

The economist studies these relations of exchange and tries to understand the *rate* of exchange: the determination of the value of commodities.

However, the same economic activity means different things to different economists, depending on what they believe is the motivation behind exchange. Is it consumers' demand, reflecting either pleasure or need, that defines 'value'; or perhaps it is the exigencies and difficulties of the production process that are determinant; or is people's economic experience as producers and consumers constrained by who controls economic resources? That is, is the consumer, the producer, or the citizen the 'dynamic' of economic activity?

Each of these conceptions of the economic dynamic implies a distinct definition of value. There are different theories of value, and in Chapters 3, 4 and 5 we shall address these different theoretical perspectives in detail.

Different perspectives on value are what Heilbroner and Milberg refer to as 'visions' (see above), and we shall see in Part IV how and why different economic perspectives imply particular 'political hopes and fears, social stereotypes and value judgements', but our concern in this chapter is to make explicit the vision of economic reality constructed by different theories of value. Remember the admonition from Daniel Bell and Irving Kristol above, that 'controversies in economic theory involve a clash of assumptions about economic reality.'

Even though economists address the same experience, because they work within the intellectual parameters of different theories of value, which define economic 'reality' differently, competing analyses can coherently arrive at distinct conclusions that imply different policy initiatives to address the same economic issue. And these analyses are each: *realist* – actual economic activity is addressed and evidence cited to validate theoretical conclusions; *rationalist* – conclusions are logically derived by analysing experience from the point of view of plausible assumptions; *activist* – policy prescriptions commensurable with accepted moral codes are advised (see Cole *et al.* 1991, Chapter 1, and Cole 1995, Chapters 1 and 6).

> Each [theory of value] points to an area of experience as matching its description, carefully displays the logic of its arguments, and arrives at appropriate policy conclusions.
>
> (Cole *et al.* 1991: 5)

As we shall see in Part IV, alternative perspectives define 'relevant knowledge' differently, and justify conclusions by appeal to a distinct concept of science. Such methodological and epistemological assumptions make for confusion between economists, if they are unaware of the implicit, intellectual parameters of economic analyses. The *same* word is a different *concept* when situated in a different perspective: words such as price, value, inflation, and economic development have a distinct significance for economists working within different theories of value. The same word has a different logical status in the argument, and there are different intellectual terms of reference between perspectives (Alford and Friedland 1985, Chapter 1). There are distinct conceptualizations of economic reality: hence in

> controversies between neoclassicists, Keynesians and institutionalists . . . [v]ery little intelligent discussion occurs . . . Sometimes there are debates, in which the two sides talk past each other . . . [each] convinced that all the evidence confirms their theory, and their opponents are equally convinced of the opposite . . . produc[ing] *distorted communication*, allowing some concepts to be communicated but blocking and distorting others.
>
> (Diesing 1982: 12, 13 and 5, emphasis added)

For an analysis of the Keynesian/monetarist controversy over the theory of inflation where 'the two sides talk past each other', see Wilber (1979). For other comparable debates in economic theory, see: Diesing (1982), Chapters 1, 11, 13, 14; Ward (1979), Book 1 Part 1, Book 2 Part 1, Book 3 Part 1; Coates and Hillard (1985); Coates (1994); Cole (1993).

Returning to Heilbroner's and Milberg's 'visions':

> the reasons for theory selection must be sought elsewhere than in empirical verification . . . theories are chosen because they yield policy implications that are compatible with one's *vision* of the economic process.
>
> (Wilber 1979: 978, emphasis added)

The economic vision defines the assumptions that underlie the analysis: assumptions highlight particular aspects of experience as being important, and specify what 'we *believe* we know'. Certainly theories have to be verified with facts, but we choose those facts that accord with our vision of economic activity and hence which plausibly explain experience.

> Let me first dispose of the question whether any economic theory is, or can be, 'correct'. Students often ask me which theory is right? This is an inappropriate question because there is no objective way of assessing whether any theoretical school is right . . . the main ones are self-contained systems, perfectly logical on their own premises . . . Empirical tests are not very relevant . . . because the objectives . . . are derived from the theories . . . The crucial questions are: whose interests does a theory serve? How does it serve them?
>
> (Seers 1983: 33)

We shall return to the issue of the interests served by theoretical interpretations in Part IV of this book. At this point in the argument it is sufficient to note that different conceptualizations of the economic dynamic explain *why* people relate through commodity exchange, *how* the rate of exchange is determined, and *what* is exchanged. All of these questions focus on the determination of economic activity: what causes value. And the causation of economic phenomena is a question of human motivation, which we can only infer from observing actual, empirically verifiable behaviour. Causation – why people behave in particular ways – cannot be directly perceived; it is an interpretation based upon the effects of action: 'in reality, what reasoned [economic] debate does take place is about *means* not ends' (Lal 1983: xi, emphasis added). The *end* of economic activity implies an interpretation and understanding of the *purpose*.

> The state of confusion in [economic] policy, and indeed amongst academic theorists, stems from a sharp disagreement over the real *causes* of . . . phenomena, and the ways they can best be addressed.
>
> (Colclough 1982: 490, emphasis added)

Economic perspectives and schools of economic thought

An economic vision defines the interpretation of economic activity. The way human, economic behaviour is understood: the *meaning* of economic activity – and hence the reason why the exchange of commodities occurs, and what is the significance of the rate of exchange, and therefore of price. This vision defines the intellectual parameters of the interpretation of economic activity, the perspective, the *theory of value*.

When economists bring that perspective to bear on actual economic behaviour in the analysis of an economic issue or problem, then a school of economic thought is defined. Thus, on any one issue, such as inflation, there are a number of schools of thought: for instance, in this case, monetarism and Keynesianism. But because they have evolved out of different perspectives, even though the same experience is being addressed, the same words mean different concepts, and 'very little meaningful debate takes place'. Indeed, economists in different schools, dealing with different issues, but part of the same perspective, have much more to say to each other. They speak the same language. So, monetarist (inflation) theorists can talk to supply-side (employment) analysts, and comparative advantage (international trade) specialists: but they have very little to say to Keynesian (inflation) theorists (Wilber 1979).

At any one time the burning economic issues of the day (politically defined) are addressed by competing theories of value, giving rise to a number of incommensurable schools of thought on the issue. Figure 2.1 shows the evolution of perspectives through the work of particular economists, at different times, studying topical issues (see also Cole *et al.* 1991: 14, and Cole 1995: 37). The period before Adam Smith does not show the same consistency as after the nineteenth century, because this was a time when people's lives were fundamentally changing as productivity and economic specialization increased, resulting in an extended division of labour in

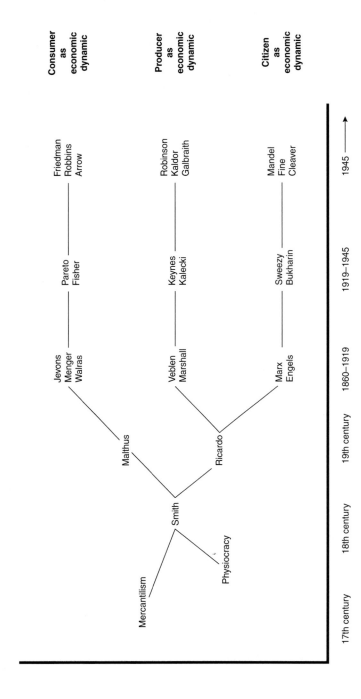

Figure 2.1 Historical evolution of perspectives on economic analysis

production. This process went hand in hand with the collapse of the feudal state, and Smith was puzzled how, given the relaxation of the political and religious control of people's lives, society and the economy was apparently able to hold itself together, even though individuals and businesses pursued their own economic agendas. Why didn't society sink into anarchy?

In 1776, in the *Wealth of Nations* (Smith 1970), Smith concluded that the motive of self-interest was sufficient to create a self-regulating economy, as if individuals were guided by an 'invisible hand'. 'It is not from the benevolence of the butcher, the brewer, or the baker that we expect our dinner, but from regard to their self-interest' (Smith, quoted Heilbroner 1992: 55). But Smith's analysis of self-interest is contradictory (Cole *et al.* 1991, Chapter 2). He is confused as to whether or not it is our self-interest as consumers, producers, or citizens, that is decisive. And since 1776 distorted communication between economists working with different interpretations of self-interest has served to further muddy the theoretical water.

There is not space here to explore the history of economic thought further, and though there is no substitute for reading these works as close as possible to the original, for contemporary commentaries see: Cole *et al.* 1991, Chapter 2; Gray and Thompson (1980); Roll (1973); Routh (1975); and the much underrated Stark (1944).

Supply and demand, demand and supply, or demand with supply

To illustrate distorted communication between economists, and how this can radically affect policy advice to decision makers, we shall look at perhaps the most basic of economic concepts, the theory of price formation as a consequence of the forces of supply and demand: that 'fundamental tool of economic analysis' (Baumol and Blinder 1991: 50).

All economists agree that, typically, as prices fall (rise) individuals tend to demand more (less) of a commodity. And as prices fall (rise) producers supply less (more) of any given commodity. But economists working within the parameters of different theories of value will have different explanations for *how* consumers' demand and producers' supply affect the price and quantity traded. The problem is that price and quantity, and supply and demand, change simultaneously. Consequently, to understand how and why demand and supply interact, and to explain the price and quantity traded, we have to interpret people's actual behaviour to imply what is the dynamic of economic activity.

For some economists, essentially it is the consumer who is determinant. Individuals buy commodities according to the pleasure and usefulness, the *utility*, of the commodity. And the greater the utility, the higher the price they will be prepared to pay: '*value depends entirely upon utility*' (Jevons 1970: 77, emphasis in original). At the price consumers are prepared to pay, producers have to decide how much they can profitably produce. Demand determines price; supply determines quantity. As an approach to economic theory, the 'consumer as economic dynamic' is addressed in detail in Chapter 3.

But:

> in reply to Jevons a catena rather less untrue than his can be made by *inverting* his order and saying:- Utility determines the amount that has to be supplied. The amount that has to be supplied determines the cost of production. Cost of production *determines* value because it determines the supply price which is required to make the *producers* keep to their work.
>
> (Marshall 1947: 818–9, emphasis added)

So, for Alfred Marshall, writing at the turn of century, and slightly after Wilfred Jevons, 'the amount that has to be supplied determines cost of production, which determines price'. At that price, a consequence of the production process, consumers decide how much they can afford to purchase. Supply determines price; demand determines quantity. This is the 'producer as economic dynamic', analysed in Chapter 4.

For Jevons, consumers' preferences will define the price, and suppliers' profitability the quantity, until the amount demanded is brought into equilibrium with the quantity supplied. But for Marshall, the productive process will determine the price of the commodity, and at that price consumers' preferences will define the quantity traded. And again, the amount traded equates supply and demand.

However, for Karl Marx:

> The real difficulty in formulating the general definition of supply and demand is that it seems to take on the appearance of a tautology . . . If supply equals demand they cease to act, and for this very reason commodities are sold at their market-values . . . [But] if supply and demand balance one another, they cease to explain anything, do not affect market values, and therefore leave us so much more in the dark about the reasons why the market-value is expressed in just this sum of money and no other. It is evident that the real inner laws of capitalist production cannot be explained by the interaction of supply and demand.
>
> (Marx 1972a: 186, 189)

While changes in the amount supplied/demanded might explain *changes* in price/quantity, and the trend towards equating demand and supply, at the (hypothetical) point where the amount demanded/supplied is equal, demand and supply no longer exert pressure for consumers/producers to change the transaction price. So what determines the equilibrium price and quantity? For Marx, the economic dynamic lies behind the appearance of economic activity: buying and selling, demand and supply. Price is an index of *power*. Certainly the price will reflect the want or need (the utility) of the commodity.

But the price will *also* reflect the ability to produce the commodity, which will depend upon technical knowledge, and also the control of resources, including knowledge, by certain people.

The economic dynamic is neither the consumer or the producer; each depends upon the other. There is a dialectical relation of interdependence. Each is the cause and precondition for the other. This is the 'citizen as economic dynamic', the subject of Chapter 5.

If value reflects consumers' preferences, then economic prosperity is a consequence of an expansion in consumption. And because consumers' choices are subjective, to maximize consumption there must be free markets where individuals can buy and sell according to their unique tastes. Ideally there should be no state intervention in economic activity (see Chapter 3).

But if it is the exigencies of production that define value and price, then because individuals' productive activity is defined by a shared technical division of labour, markets are now *systems* through which individuals producers cooperate, not *mechanisms* that function to maximize individual consumers' utility through competitive exchange. And where the producer is assumed (believed) to be the economic dynamic, the economy has to be managed by experts, economists, to create an environment conducive to technical cooperation between individuals in the general interest. A key theme is the full employment of economic resources (see Chapter 4).

However, for Karl Marx, the 'inner laws of capitalist production' have a logic deeper than market exchange and the interaction of the forces of supply and demand. People's economic activity as both consumers and producers is constrained by the minority of people that control economic resources. The dynamic of the economy reflects the power of citizens (who are both consumers and producers). And as people react and respond to the frustrations of market exchange, frustrations that deny some people the fulfilment of their economic potentials – for instance, they might become unemployed – then, in some way, the power of the individuals who control economic, social, resources will be challenged, perhaps leading to social change.

The concept of economic equilibrium (and perfect competition) is a myth. It is not just that it is unrealistic, economies being characterized by imperfect competition, but, given people's motivation to fulfil their changing potentials, as a concept it is logically impossible. Rather, it should be the purpose of economic theory to explain *change*, as a consequence of the injustice perpetrated by the control of the social means of production by individual capitalists. The economic activity of the capitalist mode of production is orientated towards the realization of profit, not the satisfaction of need, and contradictory class interests explain the evolution of economic activity (see Chapter 5).

The elementary and the complex

In economics there is nothing more elementary than the most complex of mathematical formulae, nor anything more complex that the most elementary concepts. While the former is wholly without content, the latter contains, latent within it, the system of economic relations in its entirety.

(Levine 1977: ix)

Relations of exchange between consumers' demand and producers' supply are the most elementary concepts in economics. But as will be argued in this section of the book, if we assume – that is, *believe* – that either the consumer, the producer, or the citizen is determinant in relations of exchange, then we define the system of economic relations. Analysis, by assumption, is orientated towards either a freely competitive economy in individuals' interests, a managed economy in the social interest, or the struggle for economic justice in the class interest of the disadvantaged.

We shall look at these conceptions of economic reality in detail in the next three chapters.

3 The consumer as economic dynamic

The subjective preference theory of value

The subjective preference theory of value

> Economists proceed on the *article of faith* that the economy is a structure that exists for the benefit of the individuals in it as human beings and, therefore, for *consumers*.
>
> (Lancaster 1974: 216, emphasis added)

In explaining relations of exchange between consumers and producers, in societies characterized by a technical division of labour, the dominant perspective for much of the period since the 1870s has been the *subjective preference theory of value* (see Cole Cameron and Edwards 1991, Chapters 3 and 4, and Cole 1995, Chapter 3). For economists working within these intellectual parameters, consumers are assumed to determine the value of commodities exchanged. Not only is this assumption a belief as to the motivation of individuals involved in economic exchange, it is also a belief that independent individuals determine the terms of exchange – an 'article of faith' that individuals as consumers are the dynamic of economic activity.

Hence:

> ... general principles can be derived from *simple* statements about how *individuals* behave in various circumstances ... we examine economic questions by first looking at *individual* decisions.
>
> (Craven 1984: 6–7, emphasis added)

These simple statements about individuals' decisions are a particular interpretation of human behaviour, based upon an (implicit) conception of human nature.

> [T]he nature of economic analysis ... consists of deductions from a series of postulates ... which are universal facts of experience present *whenever* human activity has an economic aspect ...
>
> (Robbins 1984: 99–100, emphasis added)

Human nature is 'obvious', we have only to look at ourselves to realize what we are. This common-sense view of the world is important in understanding the ideological power of this perspective. The social status quo in market-based societies, which are characterized by relations of competitive exchange, is justified. Apart from offering a superficially plausible account of economic behaviour and inequality, the theoretical conceptions of the subjective preference theory of value are also legitimated by an ideology of science: positivism – an investigative methodology that justifies particular economic interests in the social control of human existence (see Chapters 15 and 18).

It is also important to note the word 'whenever', emphasized in the quote from Robbins above. Because human beings everywhere are assumed to have the same nature, policy prescriptions of the subjective preference theory of value are considered to be applicable anywhere in the world: Nigeria, Nicaragua, Norway or Nepal. These policy prescriptions are seemingly consistent with human nature: a society organized on these lines is 'natural'. There is *no* alternative. This implication will be important when we consider the perspective on development, modernization theory, which is congruent with the subjective preference theory of value (see Chapter 11).

These 'universal facts of experience' are:

1   Every individual is different. We are all endowed (either by a God or by genetic inheritance) with a unique set of tastes and talents, which define our preferences. Such preferences determine our behaviour in general, and our economic activity in particular. Because these characteristics are an endowment, they are *not* a product of society, and essentially human behaviour and economic motivation is natural, and is independent of society.

2   Individuals like to enjoy themselves – what economists call *maximising utility*. But because preferences are subjective, reflecting the individual's endowment of tastes and talents, the only way these preferences are revealed is through the commodities that individuals choose to buy and sell (the theory of *revealed preference*). Hence free markets are a necessary corollary for the maximization of utility. Since humans are independent of society, there is no 'social' rationale to human existence. The only point to life is for individuals to fulfil their own interests and maximize utility. Now they may *choose* to be altruistic, and share their assets and cooperate, but this should be their choice, and not a canon of social policy to be imposed by political authority.

3   Individuals suffer from a condition known to economists as *diminishing marginal utility*. The more you consume of any one commodity, after a certain amount, each extra (or marginal) commodity that is consumed will give less pleasure than the last (utility will diminish). Hence people prefer to consume a range of commodities, and their relative preferences will be expressed by the prices paid for alternative commodities in the free market. Since the utility enjoyed from any one source diminishes as consumption increases, if total utility is to be maximized the utility from the last unit – the marginal unit – of each source consumed should be the same. The complex mathematical conditions for satisfying this

condition were specified at the end of the nineteenth century by the *marginalist* school of economic thought (Blaug 1968, Chapter 8). Hence to maximize utility, individuals must consume a range of commodities – a range that will differ between individuals reflecting their subjective preferences as revealed by the commodities they choose to consume.

Now, everyone I know is different, likes to be happy (even masochists enjoy being hurt), and prefers to choose between a range of alternatives. These observations really seem to be 'universal facts of experience'. But here is where assumptions enter the story in a logical sleight of hand. It is believed, it is an article of faith, that these universal facts of experience determine prices (which are the same as values). This is not a 'fact', it is an assumption. That prices are determined by people's differences, their motivation to be happy, and their predilection for choice, does not logically follow from these universal facts of experience.

Given this article of faith, markets are understood to allocate available resources, to produce those commodities that give consumers the most pleasure: those that maximize their utility.

> Economics is the science which studies human behaviour as a relationship between ends and *scarce* means which have *alternative* uses.
>
> (Robbins 1984: 16, emphasis added)

Scarcity is defined relative to the preferences of consumers, who *always* want more.

There are two broad emphases within the subjective preference theory of value. *Neoclassical* economic theorists, working within the intellectual parameters of the conceptions of human nature outlined above – the universal facts of experience – construct a model of utopia: a society in which individuals could behave according to their (assumed) nature; a society based on self-interest, and yet a society in which individuals would have an incentive to cooperate to specialize, augmenting economic efficiency and leading to economic growth. It is a model of *perfect competition* (see below), and economic policy is orientated towards making the real (imperfect) world more and more like the ideal (perfect) society.

By contrast, *Austrian* economic theorists concentrate on how economies actually work, rather than on how they ought to work.

Austrian economic theory

> [M]ainstream [neoclassical] theory fails to explain how markets do in fact *come* to work. It explains in great detail the relationships that would prevail in markets that already do work [the model of 'perfect competition'] . . .
>
> (Kirzner 1997: 13, emphasis in original)

Carl Menger, who published his *Principles of Economics* (Menger 1981) (originally in German) in 1871, is considered to be the founder of this school of thought. In recent times Ludwig von Mises (Mises 1960) and Friedrich Hayek (Hayek 1940, 1978) have developed Austrian thinking, applying this approach to contemporary economic issues (on the evolution of Austrian economic theory, see Kirzner 1994a, 1994b, 1994c).

Mises' contribution has been to emphasize the role of entrepreneurs in discovering profitable lines of production.

> The success which capitalist market economies display is . . . for less efficient, less imaginative courses of productive action, to be replaced by newly *discovered* superior ways of serving consumers . . .
>
> (Kirzner 1997: 31, emphasis added)

On the behaviour of entrepreneurs discovering profitable opportunities in the context of business cycles, see Cowen (1997).

We are *all* entrepreneurs in one way or another; we all choose between alternatives to maximize personal gain. Entrepreneurial activity is essentially a question of alertness to economic opportunities that 'have *until now been perceived by no one at all*' (Kirzner 1985: 29, emphasis in original). However, some people are particularly adept at recognizing economic opportunities, and at organizing and deploying resources to satisfy an unmet need, or to fulfil an existing need more efficiently. The reward for this talent is profit. But entrepreneurship is costless; it is economic activity that realizes a '*pure gain*' (Kirzner 1985: 28, emphasis in original). Profits are the result of creative alertness; they are *not* a share of previously existing value. Friedrich Hayek has added to this insight by emphasizing that the market is essentially a process which disseminates information.

> [T]he existence of systematic market forces means the existence of a *spontaneous process of learning*.
>
> (Kirzner 1992: 201, emphasis added)

The future is undetermined and intrinsically unknowable, being the result of a myriad of individuals' unique subjective perceptions and preferences, or what Hayek calls 'sense qualities', 'which most of us are inclined to regard as the *ultimate* reality' (Hayek 1942: 31). There is not a social reality beyond that conceptualized in individual activity (Lawson 1997, Chapter 10).

> [There] can only be 'error' if the future *can* be known. But if the future is acknowledged to be created by choices that are yet to be made, how can it be known.
>
> (Buchanan and Vanberg 1990: 27, emphasis in original)

Consequently, there can then be no coherent notion of a social or an economic optimum, and hence the technique of cost–benefit analysis addressed in Part II of this book, intended to manage economic affairs to create an equilibrium and hence environmentally sustainable development, is an irrelevance. Free markets do lead to an 'optimum', but this cannot be defined independently of the working of *actual* markets, which reflect the preferences of innumerable individuals: relevant information is scattered among many minds and cannot be 'theoretically' discovered by the mind of the economist.

> The social advantages [of market processes] thus achieved do not constitute 'social optimality' . . . They constitute instead a coordinative process *during which* market participants become aware of mutually beneficial opportunities for trade . . .
>
> (Kirzner 1997: 67, emphasis added)

This problem, of defining a social optimum, independently of market exchange, and yet assuming that economic exchange reflects individuals' subjective preferences, constitutes a contradiction within the beliefs and assumptions of neoclassical theory, to which we shall now turn.

The neoclassical utopia

Building upon the universal facts of experience – that individuals are different, with distinct subjective preferences; prefer to enjoy themselves and maximize utility; and like to choose between alternative sources of utility – neoclassical theory goes on to argue that the *only knowledge* needed by individuals to maximize their pleasure – the *sole purpose* of economic activity, 'to maximize pleasure is the problem of economics' (Jevons 1970: 101) – is the relative market prices of commodities.

Within neoclassical theory there are a number of schools of thought, such as monetarism, new classical macroeconomics, and rational expectations theory, (Cole 1995: 55, and Chapter 2 above), which share the same assumptions as to the nature of the exchange relation between consumers and producers (Cole 1995, Chapter 3). Neoclassical theorists understand that prices reflect competitive bargaining between producers and consumers over the rate of exchange of commodities – a rate of exchange that appears as a price. The final market price is thought to be determined by the utility (pleasure) enjoyed in the consumption of the commodity, indicated by what consumers are prepared to pay. At this price producers decide on the quantity that can be supplied profitably.

However, for some neoclassical theorists, the *Marshallians*, after the economist Alfred Marshall (1842–1924), it is the *production process* that sets the value of commodities. And consumers merely decide how much they can afford to purchase, and therefore the quantity exchanged (see the discussion on the theory of supply and demand in Chapter 6). Alfred Marshall's thought subsequently developed into the *cost-of-production theory of value* (see Chapter 4). The alternative neoclassical interpretation of economic activity, a development of the thought of Wilfred Jevons (1835–82) and Leon Walras (1834–1910), understands consumers to be the important decision maker with regard to the determination of value (Cole 1995: 41–8). On the development of economic thought into competing perspectives see Chapter 2.

As we shall see below, Jevonian neoclassical economic theory (hereafter just referred to as 'neoclassical theory'), building on a conception of human motivation, and thus an understanding of human nature, constructs a model of what utopia would be like: an image of society in which individuals would be able to behave according to their nature. Such a society would be characterized by a perfectly competitive economy. This model of perfect competition provides the template for the design of neoclassical, economic policy initiatives: policies that are intended to make the real world more like the ideal.

Given the assumptions and beliefs that underlie neoclassical theory, the *indifference curve* analysis addressing consumer choice, and the *isoquant analysis* addressing producers' decisions, show that if market prices are a product of competitive exchange in free markets then there will be general equilibrium. Everybody who is party to the processes of exchange will maximize their utility given their unique endowment of tastes and talents (Cole 1995: 58–60). In the sphere of consumption, according to their tastes, individuals enjoy utility; in the sphere of production, according to their talents, individuals suffer *disutility*. And the whole market process is a mechanism through which the pain of production is offset by the pleasure of consumption.

> The theory . . . is entirely based on the calculus of pleasure and pain; and the object of economics is to . . . [purchase] pleasure . . . at the lowest cost of pain.
>
> (Jevons 1970: 91)

Crucially, the link between the spheres of consumption and production is the *entrepreneur*. Market prices for commodities indicate the strength of demand: an index of *pleasure*. The prices of productive inputs, or *factors of production*, indicate the displeasure that individuals are willing to tolerate to produce: an index of *pain*. Entrepreneurs balance utility (pleasure) and disutility (pain): they coordinate the factors of production so as to produce the commodities which consumers desire. And the reward for this particular talent is profit.

But now, in contradistinction to Austrian theorists, entrepreneurial activity is *not* costless. For neoclassical theory, exchange is a balancing act between pleasure and pain. In this context every human activity has an *opportunity cost*. By consuming a bar of chocolate I am not consuming a pint of beer

(I cannot afford to consume both!). So the real cost, the opportunity cost, of the chocolate is the pleasure I have forgone by not drinking the beer. And I can only consume chocolate or beer if I bother to get out of bed in the morning and go to work and/or not enjoy my time in pursuing leisure activities – going fishing. Now, perhaps the entrepreneur *could* have been a brain surgeon. And so, by occupying him/herself in organizing production to the benefit of consumers, s/he is not enjoying the income that they would have received had they become a brain surgeon.

There is a real opportunity cost to being an entrepreneur, and we shall never know what they *could* have been, so we can never say what their level of income, profit, *should* be. What entrepreneurs are worth is as long as a piece of string! But, in addition, as long as free markets obtain, the fact that particular individuals are inordinately rich is a clear sign that these particularly talented and hard-working individuals give a great deal of pleasure to consumers. Because tastes and talents are unique to individuals, whose preferences are subjective, the subjective preference theory of value, beyond recommending free markets that will ensure that people receive income commensurate to their value to society (consumer sovereignty), has nothing to say about economic inequality and the distribution of income.

The operation of a market society makes it impossible to define a just distribution of income, or a social optimum for economic activity. Because the real cost of economic activity is the opportunity cost of alternatives forgone, the same change in prices will have a different significance for different individuals. The opportunity cost of economic activity reflects the unique subjective preferences of individuals, and the effect of a change in price will be different for each individual. Therefore an optimal set of prices cannot be defined.

> Given the initial distribution of resources, each individual entering the market may be conceived to have a scale of relative valuations; and the interplay of the market serves to bring these individual scales and the market scale as expressed in relative prices into harmony with one another. Prices, therefore, express in money a grading of the various goods and services coming onto the market. Any given price, therefore, has significance only in relation to the other prices prevailing at that time. *Taken by itself it means nothing.*
>
> (Robbins 1984: 55, emphasis added)

Perfect competition

It can be demonstrated, using the an analysis called the *Edgeworth–Bowley Box*, that in market transactions a set of prices exists such that both parties can be better off – their utility, and their enjoyment is increased (Cole 1995: 60). Under conditions of general equilibrium every transaction must be voluntary, reflecting individuals' subjective preferences, and hence will be to the advantage of each party to the exchange. And all of each individual's income must be purposefully used (spent or saved): individuals do not suffer the disutilty of production for no purpose, but to enjoy themselves (or feel more secure).

In economists' jargon: for general equilibrium to obtain, the excess demand for each good traded must be equal to zero. Market supply must equal market demand. And it can be shown, theoretically, that a particular price configuration exists that will precisely cancel out excess supply and excess demand for all traded goods and factors.

So, in general equilibrium, no one produces commodities that are not sold for a price that compensates them for the disutility of production, and any individual can purchase as many commodities as they want, only subject to their income and their tastes. And individuals' incomes reflects their talents and ability for satisfying other individuals' subjective preferences for pleasure: what consumers are willing to pay for their economic activity. This is *consumer sovereignty*. The consumer is the economic dynamic. Each individual is in equilibrium with every other individual: so no one can become better off without someone else becoming worse off – a condition known as *Pareto optimality* (after the economist Vilfredo Pareto, 1848–1923).

> [W]here decision makers so allocate the resources under their control that there is no alternative allocation such that any one decision maker could have his expected utility increased without a reduction occurring in the expected utility of at least one other decision maker.
>
> (Reder 1982: 11)

A Pareto *improvement* is a movement towards such an ideal state: an ideal that *only* exists in theory, and cannot be empirically verified, since satisfaction is a subjective trait. Such a state of general equilibrium, characterized by Pareto optimality, is achieved under conditions of *perfect competition*.

The concept of Pareto optimality refers only to the maximization of individuals' utility. An economy in which there were a few very rich individuals and a majority of very poor people would still be 'optimal' if, by making the poor richer, the rich were made poorer! Remember, because we have assumed that individuals' tastes and talents are an endowment, and that activity is a consequence of subjective preferences, we cannot, logically, even ask any questions about society, only about individuals' freedom of choice. This conclusion helps to explain why the subjective preference theory of value is ideologically so powerful. The social status quo, in market societies, is legitimated and justified (see Part IV and Chapter 17). Indeed, any maldistribution because of market imperfections – and markets are *always* imperfect – should not be compensated by a redistribution of the 'ill-gotten' gains of the rich to the poor, because this will only make the 'imperfections' even worse! The *only* policy consistent with the principle of Pareto optimality is that any imperfections be ameliorated by government policy so that the distortions in the future are less than they have been in the past. A step towards perfect competition is called for.

For the neoclassical economist:

> a tax/subsidy system based on *income* difference which aimed at legislating for a desired income distribution ... would affect the choices individuals make at the margin between work and leisure. By distorting the initial, *ex hypothesi*, efficient allocation, the income-based tax/subsidy system, though improving the distribution of income, would impair the productive efficiency of the economy.
>
> (Lal 1983: 14, emphasis in original)

That the rich should hold on to any gains that have been made because of market imperfections is 'scientifically' sanctioned. But as we shall see below, when discussing the 'economics of the second best', there is *no* theoretical basis for such a common sense response to inequality.

Modern welfare economics has defined in great detail the conditions under which a perfectly competitive market economy satisfies the equilibrium conditions for Pareto optimality. And these ascribed *social-efficiency* properties are generally treated as a reasonable approximation of actual, capitalist economies. Imperfections in the market, which result in a less than optimal economy, are identified as those features of an economy that violate the necessary conditions for perfect competition.

> The Utopian theoretical construct of perfect competition ... becomes relevant as a reference point by which to judge the health of an economy, as well as the remedies suggested for its amelioration.
>
> (Lal 1983: 15)

The conditions of perfect competition are:

1 In any market transaction there are so many buyers and sellers that no one individual, however much they buy or sell, can affect the price. Hence similar commodities in the same market are always bought and sold at the same price: this is, of course, *impossible*.
2 Markets are characterized by lots of small firms who are all in competition, and the costs of production are the same whatever the output. There are no economies of scale (the cost of production, and the selling price, declines the greater the number of commodities produced): this is, of course, *impossible*.
3 Every transactor has perfect knowledge so that everyone is able to buy (sell) in the cheapest (dearest) markets: this is, of course, *impossible*.
4 All individuals have the freedom to be a producer and enter any market: this is, of course, *impossible*.

In a famous passage in *Alice in Wonderland* [in fact it is in *Through the Looking Glass*] the Red Queen declares that she often thinks of six impossible things before breakfast. The model of competitive equilibrium [perfect competition] appears to be requiring us to move into this kind of world.

(Ormerod 1994: 71)

But the model of perfect competition cannot be dismissed for being unrealistic. It is not meant to describe any 'real' economy: 'everyone can see that we are not dealing with any actual economy' (Hahn 1994: 246). If the universal facts of experience are believed, *and* it is assumed that these characteristics determine prices, then the model of perfect competition is the image of Utopia: a society where individuals are able to behave according to their nature. Economic policy is then intended to make the real world more like the ideal, and it is common sense that such a step, any Pareto improvement, would be in the right direction, adding to aggregate utility.

Unfortunately, common sense is not always a good guide ... forty years ago an important contribution to the literature demonstrated that serious problems exist for the model if *any* of its assumptions are breached.

(Ormerod 1994: 83, 82)

This is the problem of the *second best*. Even if the conditions of perfect competition *did* apply, which logically they cannot, there would be no way in which this could be known. And in the absence of any one of these conditions, any policy that is intended to make the economy marginally more competitive – say, by restricting the monopoly control of markets by transnational corporations, or by limiting the power of trades unions to protect their members' wage levels, or by privatizing public utilities such as the water industry to engender greater competition – may, in the aggregate, *reduce* utility enjoyed (Lancaster and Lipsey 1956). Any potential improvement is an empirical question, and since utility is defined by individuals' subjective preferences, such an inquiry faces insuperable methodological difficulties. The belief that free markets generate a net increase in the utility enjoyed by consumers is just that – a *belief.*

For *Chicago economists*, a particular school of neoclassicism, imperfect competition,

... market failure and government intervention are taken to be sufficiently infrequent and have a sufficiently limited impact that the hypothesis ... of perfect competition provides a good approximation to the way markets work.

(Backhouse 1998: 26)

Such a position clearly contradicts Lipsey and Lancaster's argument about the 'second best'. Again, it is a question of belief.

A *social optimum* defined by economists may or may not be the best use of economic resources, but it certainly is not an unambiguous improvement.

Neoclassical economic policy

Given that, for individuals, the purpose of economic activity is to maximize personal utility, it is always in their interests to distort relations of exchange to their favour. Anyway, no individual would actually know that they were exchanging in a perfectly competitive economy even if this were to be the case. Such an imperfect personal advantage might be achieved under cartels between producers and monopoly price agreements, by restricting entry into markets, by not divulging market-sensitive information, by trades union activity to protect livelihoods, and so on. Hence a strong state is needed to enforce conditions of free exchange and impose law and order, and to restrict the poorer, less talented members of society from pressuring the more talented from enjoying their just rewards.

Economists working within the intellectual parameters of the subjective preference theory of value believe that consumers' demand determines the value and hence the price of commodities, and that markets must be free to allow individuals to reveal their particular, unique subjective preferences. Individuals are independent of society, their preferences (and prices) reflecting their endowment of tastes and talents.

> There is no such thing as society. There are individual men and women and there are families.
>
> (Margaret Thatcher, then British Conservative Prime Minister, quoted in the *Observer* 27 December 1987)

A nation is *not* one large decision maker, but is composed of a complex collection of decision makers, who are independent of society.

> The free market argument . . . [holds] that all social action must be sanctioned by the will of rational individuals composing society. Under this view society is nothing more than the aggregate of the individuals composing it . . . norms should not be imposed on society by a government . . . [and] people have an inviolable right to keep what they have earned.
>
> (Schotter 1990: 1–2)

Economic problems, at root, imply *market failure*: the utopian conditions of perfect competition not obtaining, leading to sub-optimal economic performance. In the context of environmental economic policy addressed in Chapter 7 below:

environmental problems arise from market failure and [economists] define the optimum state of the environment as that which would hold were the sources of market failure corrected. From this diagnosis of the problem comes a prescription of treatment, policy instruments, such as pollution taxes and marketable permits, which *correct the market failure* and bring about the *optimum state of the environment* ... [so that] goods and services are produced efficiently and exchange at the lowest prices ...

(Bowers 1997: 33, emphasis added)

Policy prescriptions are orientated towards moving the economy towards perfect competition; market failures are a consequence of arbitrary restrictions of the forces of supply and demand. And although such failures are unknowable, the faith of neoclassical economists gives them a belief that they know how to make other people better off, even though they explicitly assume that such a Pareto improvement could *only* be perceived by individuals themselves, and even then, individuals would not be aware if others' utility had been decreased as a consequence of their advantage: a state of affairs that no self-respecting neoclassicist could endorse.

Inflation

For neoclassical theory, price inflation reflects an expansion in the supply of money.

[L]ong-continued inflation is always and everywhere a monetary phenomenon that arises from a more rapid expansion in the quantity of money than in total output – though ... the exact rate of inflation is not precisely or mechanically linked to the exact rate of monetary growth.

(Friedman 1974: 10)

Money is the means by which individuals indicate their preferences: it is 'demanded essentially for transactions purposes. The role of money is to bring buyers and sellers together with a minimum of fuss: it is the glue that links supply and demand' (Harrigan and McGregor 1991: 110). And if consumers have more money to spend, perhaps because the government to win election advantage spends more on the health service or on the provision of education, or to reduce unemployment some jobs are subsidized, or in one way or another state spending exceeds taxation, then prices will be bid up as people have more spending power for the same amount of purchases: crucially, individual consumers are not having to suffer the pain to pay for the pleasure.

There will be inflation, which subjective preference theorists believe will destabilize the economy, inevitably leading to unemployment.

The information that is important for the organization of production is primarily about *relative* prices – the price of one item compared to another. High inflation . . . drowns that information in meaningless static.

(Friedman M. and R. 1979: 37, emphasis in original)

Although the diagnosis for subjective preference theorists is obvious, the prescription is far from easy. The rate of monetary growth has to be cut, meaning the state deficit has to be cut; spending reduced or taxes increased. This will generate lower economic growth and a (theoretically temporary) increase in unemployment, without any immediate fall in the level of inflation. There is no mechanical link between the rate of monetary growth and the rate of inflation. Money and prices are mediated by consumers' subjective preferences, and there is no way of predicting, except in very broad terms, the effect of monetary restraints on economic activity. Indeed, monetarist economists cannot even agree what 'money' is.

[T]here is no single . . . definition of money. In the ten years from 1976, at least seven different definitions of money were measured and monitored in Britain.

(Smith 1987: 4)

It was Milton Friedman's ability to apparently explain the rising inflation of the 1960s (Friedman 1968), an experience that was inexplicable in terms of the then dominant Keynesian orthodoxy (see Chapter 4), that heralded the regeneration of *free market economics* after two decades of state economic intervention and management.

Employment

A related policy emphasis is employment. In the 1960s and 1970s governments were blamed by monetarist economists for trying to maintain an artificially low rate of unemployment, by subsidizing jobs: subsidies that led to an increased rate of monetary growth and ultimately to inflation.

For neoclassicists, the market for labour will naturally set wage levels, so that in conditions of perfect competition all who want work will find work. But for some individuals the wage rate will be too low to compensate them for the opportunity cost, the disutility of work, and they will choose not to work. In this sense, economists can never say what *is* the natural rate of unemployment.

The 'natural rate' is 'natural' only in the sense that actual unemployment tends to move automatically towards that rate.

(Harrigan and McGregor 1991: 121)

A level of unemployment below this level means, essentially, that some workers are being paid more than the equivalent utility they are producing: the utility of the wage packet is more than the disutility of the work. And the difference has to be paid for by consumers paying higher prices.

Hence lower unemployment is achieved by increasing the opportunity cost of *not* working. Unemployment and welfare benefits should be decreased, making unemployment more of a disutility. Secondly, employment can be increased by making it easier to employ people. If employers face lower costs in creating jobs (less requirements to provide holiday pay, sickness, maternity and pension benefits, and no right of labour to union recognition, etc. – so-called *supply-side economics*, which reduces the opportunity costs for employers in providing jobs), then employment will increase.

> [E]mphasis is placed on 'pricing the unemployed into jobs' both by reducing wage rates and by lowering the range and level of benefits to increase work incentives.
>
> (Ditch 1991: 33)

Public goods and environmental economics

A commodity is a public good when, even though it is being consumed by one individual, other people cannot be excluded from consumption. Examples include street lights, the police force, and public service broadcasting.

Thus there is a problem in the calculus of pleasure and pain. Some people are enjoying the pleasure and not suffering the pain, and in all conscience we can't have that! In the absence of market forces automatically reconciling different individuals' tastes and talents, economists have to estimate the value of the marginal utility/disutility traded, and simulate a market. This is an application of a technique known as *cost–benefit analysis*, which will be addressed in detail in Chapter 7 when we look at *environmental economics*.

The equivalent to subjective preference theory economists in the field of environmental theory are the *egocentric* theorists, who see the environment as a source of pleasure for consumers:

> the value of the natural environment [is] in its instrumental role as a means of meeting human needs and gratifying desires.
>
> (Malnes 1995: 117)

With the concern to enjoy ever more utility, pessimistically, the inexorable growth in population (and the number of consumers) will preface environmental catastrophe; or there is an optimistic faith in individuals' basic intuition and motivation to respond to 'the tremendous power of self-interest' (Block 1990: 281). And individual incentives, in the form of differential prices, will avert economic crisis, and there is a corresponding emphasis on extending individual property rights (to move towards perfect competition); or trading pollution permits as commodities to achieve an optimum level of pollution: or, in response to market failure, there is faith in economists' abilities to define an economic optimum by theoretically constructing shadow prices and perfect competition in a cost–benefit analysis – all these options, theories and policies will be addressed in detail in Chapter 7.

4 The producer as economic dynamic

The cost-of-production theory of value

The cost-of-production theory of value

For the subjective preference theory of value, individuals' tastes and talents, and their resultant preferences, define value. And because preferences are subjective, economic policy has to be oriented towards achieving free exchange between producers and consumers. But we saw in Chapter 2 that for Alfred Marshall the 'cost of production determines value because it determines the supply price which is required to make the producers keep to their work'. It is not the consumer that is the economic dynamic in relations of exchange, but the producer.

For Marshall, markets are not mechanisms by which individuals compete to maximize utility in consumption, but systems through which individuals cooperate to maximize output in the sphere of production. 'Most people think of themselves first of all not as consumers but as producers' (Galbraith and Salinger 1981: 164). Production processes imply a technical division of labour: there is technical specialization. Individuals with different technical skills cooperate to produce a finished product.

Individuals are not independent utility maximizers, but dependent producers. Individuals' economic behaviour is contingent upon others', technically defined, economic activity. The technical division of labour sets the parameters of economic activity: individuals are not free to choose. Hence markets cannot now be conceived of in terms of the utopian ideal of perfect competition, which ensures that prices reflect consumers' satisfaction. Economic analysis now has to reflect the actual technical basis of relations of exchange.

> It is easy enough to make models on stated assumptions. The difficulty is to find assumptions that are *relevant to reality*.
>
> (Robinson 1971: 141, emphasis added)

Now *time* is of the essence (time is not a problem for the perfectly competitive model of markets). Over the 'short period' (defined technically) the scale of output cannot be altered: it takes time for an investment in production to come on stream and produce commodities for exchange.

> A plant expansion programme in heavy industry . . . is a fifteen- or twenty-
> year programme. And the same is true . . . of a decision to build a new
> store or to develop a new type of insurance policy.
>
> (Drucker 1954: 89)

Consequently:

> *as a general rule*, the shorter the period we are considering, the greater
> must be the share of our attention which is given to the influence of
> demand [consumption] on value, and the longer the period, the more
> important will be the influence of cost of production on value.
>
> (Marshall 1947: 349, emphasis in original)

Competition is still important. In competitive markets producers, to survive,
have to earn profits, requiring that they adopt the most efficient, *optimal*,
technology: the technology with the lowest cost of production and the potential for the lowest price. For 'Marshall . . . prices may be indices of scarcity,
but scarcity reflects the underlying conditions of *production*' (Backhouse 1998:
153, emphasis added). And the optimal technology defines *equilibrium price*
– not the maximization of consumers' utility.

> Marshall [and cost-of-production theorists in general] did not see that any
> optimization procedure – that is any problem of combining factors [of
> production] in the 'best way' – depends essentially on a system of prices
> . . . [but] could . . . as a rule be formulated in purely *technical* terms . . .
>
> (Frisch 1950: 59, emphasis added)

Because economists are concerned to understand technical systems of
production and the markets through which individuals cooperate to produce,
the analysis focuses on the whole economy: the *macro*economy. Where the
emphasis is upon consumers' choice, with relative prices reflecting individuals' utility, the analysis focuses on the individual: the *micro*economy.
Macroeconomics is the study of the whole economy, addressing aggregate
variables, such as national income, output, inflation, unemployment and
investment. Crucially, and in contradistinction to the subjective preference
theory of value, the economy is more than the sum of the individuals: there
is a wider interest above and beyond that of individual preferences. And the
economic 'system' has to be managed to allow individuals to, as far as possible,
satisfy their subjective preferences within these social parameters. Indeed,
economic problems occur when individuals' subjective interests conflict with
the wider social interest, and the state has to intervene to ensure that this
wider interest prevails over potentially disruptive individuals' preferences.
Cost-of-production economists' analyses are invariably focused upon how the
economy might be better organized to maximize the cooperative technical
potentials of individuals' economic activity.

No doubt men . . . are capable of much more unselfish service than they generally render: and the supreme aim of the economist is to discover how this latent social asset can be developed most quickly.

(Marshall 1947: 9)

Economics is not the marginal, scientific, calculus of individuals' pleasure and pain, but is about modelling *systems* and the art of managing human economic behaviour.

Economics is a science of thinking in terms of models joined to the art of choosing models which are relevant to the contemporary world . . . Good economists are scarce because the gift of using 'vigilant observation' to choose good models, although it does not require a highly specialized intellectual technique, appears to be a very rare one.

(Keynes, quoted Moggridge 1976: 26)

Where value is determined by the cost of production there are two pre-eminent concerns of theorists: the efficient use of productive resources, and the price of those resources. The efficient *use* of resources is a technical question of using the most advanced technology, and an organizational question of fully employing available resources. The *price* of those resources is essentially a question of economic distribution. If individuals are to cooperate to produce, not only is there not a strong rationale for inequality – people are not independent actors being rewarded for their personal talents, individuals are dependent producers – but inequality is dysfunctional. Where there is a large disparity of income between the rich and the poor there is unlikely to be effective cooperation: this is a recipe for strikes and economic disruption.

The outstanding faults of the economic society in which we live are its failure to provide for *full employment* and its arbitrary and *inequitable distribution* of wealth and income.

(Keynes 1936: 372, emphasis added)

Imperfect competition

Long-run market price is the equilibrium price, when production techniques have had time to adjust to the level of consumer demand, and produce most efficiently at the lowest cost. And clearly, competition is important in pushing prices down to this level, although competition driven by individual gain tends to create unemployment and act against the broader social interest. The concern is with actual, *imperfect* markets. And to try to conceptualize how imperfect markets might function, models of *imperfect competition* were developed: Alfred Marshall's model of *natural monopoly* (Marshall 1947, Chapter XIV); Edward Chamberlain's model of *monopolistic competition*

(Chamberlain 1933); Joan Robinson's model of *imperfect competition* (Robinson 1933); and Paul Sweezy's theory of *oligopoly* (Sweezy 1939). All of these models highlight the way producers' *monopoly power* – that is, there is not the assumption of many small producers that is fundamental to perfect competition – affects prices. Prices are typically higher than the mythological perfectly competitive ideal.

Although these alternative conceptions of how markets function called into question the relevance of models of perfect competition, markets were still seen as institutions through which individuals exchange. There was no understanding of the economy as a system. These models did not posit an alternative conception of how the whole economy, the *macro*economy, might work. It was not until the work of John Maynard Keynes, and his book *The General Theory of Employment, Interest and Money* (Keynes 1936), which focused on the decisions of producers in determining macro economic activity, that there was a fully worked out alternative framework to the subjective preference theory of value and the model of general equilibrium.

> [T]he importance of . . . [Keynes'] *General Theory* lay in its presentation of both a consolidated critique of the then orthodox view and an alternative set of relationships which proved a rationale for a policy of government intervention.
>
> (Love 1991: 169)

Keynes' theory was essentially a theoretical response to a political problem. Orthodox policy prescriptions for the economic recession of the 1920s and the depression of the 1930s were intended to make economies *more* prefect. Competition was to be encouraged, forcing down prices and wages, and increasing unemployment as inefficient producers went out of business. To stave off political unrest and possible revolution governments *had* to intervene in economies by spending money to create jobs.

> The spending measures were undertaken by people who had been in opposition through the worst of the crisis, and were acutely conscious of having been brought to power by a wave of popular revulsion against the policies of deflation.
>
> (Bleaney 1985: 73)

And Keynes theoretically rationalized such policy initiatives.

Issues of distribution, eliminating extremes of wealth and poverty, and managing economies to ensure full employment and the full utilization of resources became the central focus of the 'art' of cost-of-production theory economic policy. Though Keynes wanted to preserve private enterprise – 'The central controls necessary to ensure full employment will . . . involve a large extension of the . . . functions of government . . . But there will still remain a wide field for the exercise of private initiative and responsibility' (Keynes 1936: 379–80) – so there would still be pressure to reduce costs, and prices. But private enterprise would have to be regulated. The principal problem lay with the allocation of investment finance to producers. Entrepreneurs in deciding on productive investment have to take a long view over the lifetime of the investment, estimating and comparing future costs and returns to predict potential profitability. But there is such uncertainty that enterprise is 'partly a lottery' only being maintained by the 'animal spirits' (Keynes 1936: 150, 161) of entrepreneurs – their propensity to take risks. However, investment finance is only forthcoming because investors speculate on the short-term, future value of assets; they do not consider the long-term viability of production processes, and consequently employment and the use of economic resources. Speculation is a gamble, and

> When the capital development of a country becomes the by-product of the activities of a casino, the job is likely to be ill done.
>
> (Keynes 1936: 159)

Hence the government has to intervene in the economy to maintain an atmosphere of optimism to encourage the 'animal spirits' of entrepreneurs. And interest rates have to be kept relatively low, in spite of the activity of speculative investors (who Keynes referred to as 'rentiers') – and in this context Keynes called for the 'euthanasia of the rentier' (Keynes 1936: 376) – which implies some system of state allocation of investment finance, not the allocation of resources to reflect the interests, sympathies and intuitions of private investors.

Economic management

Cost-of-production theorists do not share the faith that economies are mechanisms through which independent individuals, as *consumers*, maximize utility. Instead they have a faith that the cooperative economic activity of dependent individuals, as *producers*, can be managed in the general interest. And just as for the subjective preference theory of value, this faith rests on a conception of human nature. Individuals are not now thought to be endowed with tastes and talents, and selfishly motivated to act in their own best interests according to their subjective preferences. Rather, individuals are social animals; they adapt to society's needs; they are moulded, *socialized*, by social forces beyond their individual control. But these forces can be managed by technical 'experts': in the field of relations of exchange, by *economists*. These beliefs in human nature are rarely a consciously held theory, but are the intuitive way human behaviour is understood: the 'vision' of Heilbroner and Milberg (see Chapter 2). And Keynes was very well aware that such intuitive beliefs were not uncommon, and were the major obstacle to his ideas being accepted, and the reason why good economists with the gift of vigilant observation are so rare. 'The difficulty lies, not in the new ideas, but in escaping from the old ones, which ramify . . . into every corner of our minds' (Keynes 1936: viii). And the majority of economists' minds did not 'escape the old ideas'.

> It [Keynes' *General Theory*] is a badly written book . . . the Keynesian system stands out indistinctly, as if the author were hardly aware of its existence . . . until the appearance of the mathematical models of Meade, Lange, Hicks and Harrod, there is reason to believe that Keynes himself did not truly understand his own analysis.
>
> (Samuelson, quoted Bell 1981: 62–3)

These mathematical models came to be known as the *Hicks–Hansen neoclassical synthesis*, often referred to as the IS/LM analysis, and sometimes as the money augmented expenditure system (hereafter referred to as Keynesian economic management). Hicks's book, *Value and Capital*, 'would have been very different had I not had the [Keynes'] *General Theory* at my disposal' (Hicks 1935: 4). He proceeded to reinterpret Keynes in terms of the 'old ideas':

> under assumptions of *perfect competition* . . . [deriving] laws of market conduct . . . which deal with the reaction of the *consumer* to changes in market conditions.
>
> (Hicks 1935: 6, 23, emphasis added)

'Hicks . . . relies on the general equilibrium method, which leads him *to rule out the crucial features* of Keynes's approach . . .' (Togati 1998: 3, emphasis added). Keynes's conclusion of the need to manage productive economic activity was melded to the idea of achieving an equilibrium through market forces, with prices reflecting consumers' preferences – what Joan Robinson referred to as 'bastard keynesianism' (Robinson 1971: 90). Keynes had argued that 'the concept of the general price-level . . . [is] very unsatisfactory for the purposes of causal analysis' (Keynes 1936: 39), and nowhere does he refer to consumers' preferences being manifested as laws of market conduct: 'the *General Theory* was converted into a pastiche of ideas . . . with their mutual contradictions and inconsistencies allowed to go unresolved' (Heilbroner and Milberg 1995: 45).

In defining such an equilibrium (a *social optimum* defined in value terms):

> Considerable leeway is . . . afforded to *subjective* judgement in evaluating a whole body of econometric evidence on any given economic relationship. This leaves ample room for *intuition* and the casual empiricism of personal observation, intermingled with the influence of *political preferences* in choosing between rival hypotheses.
>
> (Levacic and Rebmann 1982: 7, emphasis added)

But even accepting the subjective, intuitive influence of political preferences in the economic definition of a social optimum, there is still the problem of the 'second best' (highlighted in Chapter 3), added to which is a logical contradiction known as the *capital controversy*.

> [T]he textbook neoclassical synthesis [the IS/LM analysis] simply juxtaposes Keynesian macroeconomics and neoclassical microeconomics in the hope that no one will bother to ask any awkward questions.
>
> (Bleaney 1985: 1)

In the 1960s the logical status of this juxtaposition *was* questioned, and awkward questions began to be asked. The debate centred on the definition of capital, and implied different theories of value.

For neoclassical theory (the subjective preference theory of value), price (value) reflects the utility enjoyed by consumers, as defined by their subjective preferences. And the price of the factors of production (and hence their earnings) is derivative of their contribution to the pleasure of consumers: a theory known as the *marginal productivity theory of distribution*. The marginal product is the marginal utility produced and enjoyed by consumers, and:

> attention is focused on the *equilibrium* set of factor prices which optimises resource allocation by maximising utility.
>
> (Burkitt 1984: 164, emphasis added)

On this theory of distribution, see: Baumol and Blinder (1991), Chapter 35; Dolan and Lindsey (1988), Chapter 28; Hardwick *et al.* (1994), Chapters 16 and 17; and Sloman (1991), Chapter 8.

The demand for, and hence the price of, factors of production is a *derived* demand, reflecting consumers' utility and hence their willingness to pay for the final product. It is believed that the wage rate is determined by the marginal utility resulting from the productive activity of labour. Though since the utility provided by a brain surgeon cannot be measured and compared to the utility produced by a taxi driver, this explanation for wage differentials and economic inequality remains an article of faith, based on a conception of human nature. But if this belief is accepted, at least the hours worked can be measured: there is something that can be defined as an amount, a supply of labour, which can be compared to a demand for labour from sick people or travellers. However, there is a logical problem with *capital*.

Capital controversy

Strictly, *profit* is the reward for the extraordinary talent of entrepreneurs in being able to organize production to meet an as yet unmet consumer demand. It is not a return to *capital*. The interest rate on investment finance is thought to reflect the utility produced by the factor of production capital. It is the price of capital. And as with any price, the interest rate reflects the supply of and the demand for capital. Hence the interest rate, the *value* of capital, moves inversely with the *quantity* of capital. But the quantity of capital is the value of capital: they are indistinguishable.

> There is ambiguity whether we are concerned with the increment of physical product . . . due to the employment of one more *physical* unity of capital, or with the increment of value due to the employment of one more *value* of capital.
>
> (Keynes 1936: 138, emphasis added)

Following on from Keynes, Joan Robinson in 1953 also asked 'how capital can be measured?' (Robinson 1953).

Capital comprises such items as word processors, airliners, warehouse space, pneumatic drills, etc., and the only way these can be aggregated into something called *capital* is to value word processors, airliners, warehouse space and pneumatic drills and add these values into one sum. Hence the quantity of capital cannot determine the value of capital: they are the same. And, theoretically, the value of capital is supposed to reflect the marginal productivity of capital (and hence the marginal utility enjoyed by the consumers of the final product, determining the demand for that particular factor of production). Consequently, the quantitative supply and demand of capital determines the price of capital: but the supply and demand of capital *is* the price of capital – these concepts are indistinguishable.

On the development of the capital controversy see Harcourt (1969, 1972).

It is the logic of the distribution of income, and hence the price of capital, being determined by the forces of supply and demand that justifies the rewards to the owners of capital: the dividends to shareholders. Market forces also legitimate inequality in general, and private enterprise and the capitalistic organization of economic activity in particular.

> Looking back with hindsight from the early 1990s we may say that the capital theory controversies of the 1950s to 1970s related not so much to the *measurement* of capital as to its *meaning* [Harcourt 1976]. Related to this perspective is the following question – What is the appropriate method with which to analyze processes occurring in capitalist economies . . .
>
> (Harcourt 1994: 29, emphasis in original)

In 1960 Pierro Sraffa's *Production of Commodities by Means of Commodities* attempted explicitly to theorize the working of an economy where the measure of capital was a reflection of the distribution of income, and hence the price of capital, and in this context was a powerful critique of the logical coherence of neoclassical theory. But nearly forty years later most economics textbooks, economics courses, and the culture and institutions of the economics profession itself ignore the logical contradictions highlighted by Sraffa, and have not been able to offer a coherent counter-argument and defence.

The 'culture and institutions of the economics profession' are synonymous with the ideological power of economic theory, which justifies and legitimates economic inequality and disadvantage.

Not only does the capital controversy highlight the logical incoherence of determining the use of economic resources by such techniques as cost–benefit analysis (see Chapter 7) – the dominant technique in environmental economics – the ubiquitous conception that the operation of the *macro*economy is based on an understanding of *micro*economic choices by consumers is also called into question since, logically, prices cannot be derived from the demands of individual consumers. This is still the basis of most courses in economics in the majority of universities in the United Kingdom, and probably elsewhere, and reflects what academics were taught when they were students in the 1950s, 60s and 70s, when the ideology of the neoclassical synthesis, Keynesian economics, was hegemonic: academics tend to teach what they were taught. Fundamental assumptions are rarely questioned; Keynes' 'old ideas', which 'ramify every corner of the economist's mind', still live on in the thinking of uncritical economists.

Hall and Taylor succinctly summarize, and continue to think in terms of, this anachronistic, contradictory orthodoxy:

> *Macroeconomics* is the branch of economics that tries to explain how and why the economy grows, fluctuates and changes over time . . . The other branch of economics is *microeconomics* – the study of the behaviour of individual consumers, firms and markets . . . macroeconomics is only as good as the microeconomics that underlies it.
>
> (Hall and Taylor 1991: 4, emphasis in original)

There is really no excuse for teachers of economics to be ignorant of, and/or ignore, such debates as the second best and the capital controversy. Even if they do not agree, as teachers it is incumbent upon them to make explicit the beliefs in terms of which individuals' economic behaviour is rationalized. The alternative is indoctrination into the economics profession: the creation of unthinking disciples, not the development of critical, thinking minds.

> [T]he hypothesis of individual rationality gives no guidance to an analysis of macro-level phenomena: the *assumption* of rationality *is not enough* to talk about social rationalities . . .
>
> (Rizvi 1994: 363, emphasis added)

It is a question of the plausibility of beliefs: that there is a logic to individuals' experience. And the confusion amongst economists

> can partly be explained by plain ignorance fomented by the poor quality
> of textbooks that avoid questions about fundamental concepts . . .
>
> (Carvalho 1992: 29)

Economic orthodoxy is essentially economic analyses based on the assumption (belief) that value (price) reflects consumers' utility maximization, and
that a utopian state of perfect competition can be conceived of, which defines
(in terms of value) the optimum allocation of economic resources. But such
an optimum can only be defined if: independent consumers are believed to
be the economic dynamic, and hence a Pareto improvement is the purpose
of economic exchange; it is believed that partial changes in market compet-
itivity imply such an improvement (ignoring the problem of the second best);
and that the contradictions of the capital controversy are an irrelevance.

The optimum solutions of economists are no more than *beliefs* that 'markets
are good for you' – a belief that justifies the economic inequalities of market
societies.

The resurgence of the cost-of-production theory of value

As we shall see in Part IV, in understanding the social parameters of individual existence – the *meaning* of human existence – the social context of
individual behaviour has to be highlighted. Implicit assumptions and beliefs
about human nature explain the social significance of individuals' intuitions
and activity, and social policy in general and economic policy in particular
are intended to construct a society in this image – a society in which individuals can behave 'naturally'.

To the extent that a plausible explanation of human activity can be based
on these beliefs, they will be accepted as a basis for the social regulation of
individuals' experience. In the heyday of Keynesian economic management
in the 1950s and 1960s it was thought that economies could be managed
by regulating aggregate (not individual) demand: too *little* effective demand
(desire backed up by income) and there would be unemployment; too *much*
effective demand and there would be inflation. Unemployment and inflation
were alternatives. This was an understanding based on Alban Phillips' inverse,
non-linear statistical relationship between the rate of change of money wages
and the percentage change in unemployment (using data between 1861 and
1957), known as the *Phillips Curve* (Phillips 1958).

> Following Phillips' paper, researchers looked for, and found, similar kinds
> of relationships in the data for the period after the Second World War,
> both in Britain and other countries.
>
> (Ormerod 1994: 119)

From the large literature on this relationship see: Begg *et al.* (1991),
Chapter 28.3; Bleaney (1985), Chapter 5; Levacic and Rebman (1982),

Chapter 18; Cole *et al.* (1991), Chapter 7.3; and Smith (1987), Chapters 2, 3 and 4.

But in the 1960s the apparently stable, inverse relationship between the rate of change of prices and the rate of change of employment broke down. No longer were unemployment and inflation *alternatives* but *complements*.

> The Keynesians could neither predict accurately what was going to happen, nor offer a convincing way out of the morass into which Western economics was sinking. The nightmare of stagflation – simultaneous rising prices and unemployment – had arrived.
>
> (Smith 1987: 45)

Economic regulation lost its rationale: 'the famous 'neoclassical synthesis' . . . foundered . . .' (Romer 1993: 5–6). However, as we saw in Chapter 3, Milton Friedman offered an explanation of stagflation, compatible with the justification of economic inequality and private enterprise (Friedman 1968). But his account called into question Alfred Marshall's interpretation of economic exchange, where the producer is the economic dynamic, and hence the basis of Keynes' economic thought. Theoretically there was a renaissance of Wilfred Jevons' balancing of the pleasure and pain according to individuals' subjective preferences.

Unemployment was not to be explained by market failure, and rectified through economic management. Rather, economies had self-righting tendencies: if allowed to be free there would be a natural tendency towards equilibrium.

There is a natural rate of unemployment, which reflects individuals' willingness to accept a reward for the disutility of work. Some people are incorrigibly lazy, and would rather live in poverty than work. Crucially, unemployment can only be reduced below this natural rate if the reward to labour (and hence prices of products) keeps rising, which gives the temporary illusion that individuals' standard of living is rising. And for this illusion to be sustained, prices have to keep rising faster than earnings. At some point prices do not rise quite fast enough, and unemployment begins to rise at the same time as price inflation.

In the late 1960s and early 1970s the notion of a natural rate of unemployment, and the related monetarist strategy for controlling inflation, provided an ideologically powerful rationale for turning the tide against Keynesian economic management: management that did not legitimate inordinate wealth, but sought to manage distribution. Henceforth, economic policy was to be a question of individual responsibility: of people competing to maximize personal utility, with success (and income) limited only by their endowment of tastes and talents, and therefore their subjective preferences. Distribution and economic inequality were not explicit concerns of economic theory. Economic policy was justified by what Edward Nell calls 'econobabble'.

> A set of clichés, phrases, and postures designed to sound important and
> appear profound, but which can easily be mastered by anyone with a
> minimal education. The need to tighten our belts, the dangers of deficits,
> the importance of sound finance, and of saving, the need to provide
> adequate incentives, the rigours of competition and the efficiency of the
> market . . .
>
> (Nell 1996: 1)

However, as we shall see in Chapter 11, the renewed emphasis on free
exchange, the ideal of neo-liberalism, failed to deliver anything that could
be described as a move towards equilibrium – an unambiguous Pareto
improvement. And while, in the 1980s, internationally neo-liberalism was in
ascendancy, by the 1990s unemployment and inequality were back on the
political agenda: job creation and poverty began to be explicit concerns of
economic policy.

In July 1993 at the Tokyo world summit President Clinton emphasized
job creation as a policy priority. At the end of that year the European
Commission published a discussion paper on unemployment in the EC. And
in March 1994 the Director General of the Confederation of British Industry
called upon the British government and private enterprise employers to address
the problems of inequality and poverty. And as I write the first draft of this
chapter the *Guardian* newspaper, referring to the forthcoming September
1998 general election in Germany, reports on Gerhard Schröder's Social
Democratic Party (SPD) campaign. At a special congress held in Leipzig:

> he showered the governing coalition with contempt for presiding over
> the worst unemployment since 1932. He promised the benchmark for
> his policy would be the battle against 5 million jobless. 'Every measure
> will be judged by whether it ensures available jobs or creates new jobs'
> he said.
>
> (*Guardian* 18 April 1998)

Gerhard Schröder subsequently won the election.

'Worldwide the intellectual political pendulum is swinging. The passion
for deregulating, tax-cutting, let-the-financial-markets-rule . . . is in decline;
the interest in the regulated market, the state-as-partner . . . is growing' (Will
Hutton, *Guardian* 6 April 1992). The restitution of inequality and job
creation to the economic policy agenda has reflected the political tensions
created by poverty and individuals' frustrations at not being able to fulfil
their social potentials (see Chapter 18). Theoretically, cost-of-production
theorists are embroiled in a scientific revolution to develop a new economic
paradigm (see Chapter 16): a model of the economic system that can plausibly explain how economies can be managed by experts, to realize the
technical potential of cooperation between individuals as producers. Much
of this debate is labelled *post-Keynesian*.

Post Keynesian theory was born of a critique of neoclassical theory in its beginnings; the label served as a portmanteau for several schools of thought that had little in common besides their rejection of neoclassical economics . . .

(Carvalho 1992: 218)

The focus is upon empirically accurate, 'realistic' analyses emphasizing productive imperatives, and central to post-Keynesian analyses are the problems of economic coordination in conditions of uncertainty (for instance, see Arestis 1988, Eichner 1987, Pheby 1989 and Lavoie 1992). In this regard, the state, which has the authority to create money, influence interest rates, encourage technical development and research through educational and regional policy, etc., can influence economic activity by creating a more predictable economic environment. And priority is placed on the regulation of aggregate demand, aid for technical research and development, regulation of personal income distribution, etc. in an institutional structure appropriate to the management and reform of economic cooperation.

But there is little theoretical unanimity within post-Keynesian economics, which is 'a community which knows what it is against, but doesn't offer anything very systematic that could be described as a positive theory' (Robert Solow, quoted Backhouse 1998: 154). Also within the cost-of-production fold the *neo-Ricardians*, following Sraffa (Sraffa 1960), emphasize the technical bases of relations of exchange, and the *New Keynesians* return to the problems of imperfect competition, with non-market clearing prices, with no tendency to move towards equilibrium (on the revivial of Keynesian thought in various theoretical guises see the *Journal of Economic Perspectives*, 1993, Vol. 7, No. 1).

But in all of these schools of thought within the perspective of the cost-of-production theory of value, the underlying conception of the nature of economic exchange, based on cooperation between dependent producers, not competition between independent consumers, means that individuals' economic behaviour has to be managed by experts: economists trained in the analysis of macro economic activity.

5 The citizen as economic dynamic
The abstract labour theory of value

The abstract labour theory of value

In contradistinction to the subjective preference and cost-of-production theories of value, it can be coherently argued that in relations of exchange it is not independent consumers, nor dependent producers who constitute the dynamic of economies, but interdependent citizens. Individuals are both consumers and producers: the behaviour of people as consumers affects their activity as producers, and vice versa. Neither sphere of economic activity is assumed to be dominant.

The subjective preferences of consumers structure the environment within which people as producers cooperate within a technical division of labour. The social context of such cooperation conditions people's experience, influencing their attitudes and personalities and ultimately consumers' subjective preferences. The social parameters of the production process, which reflects individuals' power to control the social means of production, in turn conferring social power, structure the distribution of income and hence what people can afford to buy: people's effective demand. And the purchasing power of consumers limits individuals' ability to choose, and hence effect their preferences.

Economies are not merely mechanisms by which consumers compete to maximize utility. Neither are economies simply systems to be rationally managed in the (technically defined) common interest, to fully utilize productive resources in the production process. Rather economies are *processes* of struggle between people with conflicting class interests over the utilization of economic resources and the fulfilment of people's economic potentials.

The analysis is concerned with the interaction *between* people. What individuals choose to do as consumers, or are able to do as producers is in itself unimportant: how and why their activity affects other people *is* important. Economic relations between people have not only to be explained but accounted for. The theorist has to identify what individuals do, how they do it, and what are the particular conditions of existence that make such behaviour 'rational' – why do people behave in distinct ways?

Relationships cannot be directly observed: you, the reader, and I, the author, are having a relationship *now*, but I can't 'see' it; can you? What the relationship is has to be inferred from people's behaviour. As a teacher I give lectures to students, but if I were to behave in exactly the same way in a hospital waiting area I would no doubt be thrown out, or even arrested. It is not only what people do, and how people behave, but why such behaviour is possible and even expected that is relevant. In the economy, we have to interpret people's activity in exchange, to rationalize their behaviour, to understand why in this particular historical and social context such behaviour is necessary. Why, in the late twentieth century, do most people organize their social existence according to the logic of market exchange?

Relationships are concepts: they only exist intellectually, in the mind. They are interpretations of actual behaviour; but they are none the less real for that. We really do exist in relation to other people. The analyst has to look behind the empirical evidence of actual behaviour to understand the social significance of such activity.

> [A]ll science would be superfluous if the outward *appearance* and *essence* of things directly coincided.
>
> (Marx 1972a: 817, emphasis added)

Individuals are both consumers and producers: for them as 'citizens' each sphere of economic activity conditions the other, and individuals constrain and facilitate others' economic potentials through exchange. Relations of exchange are *dialectical*. Dialectics is more than interdependence. Individuals certainly depend upon, and set the parameters to, others' economic activity. In this sense people are interdependent.

> The properties in individual human beings do not exist in isolation but arise as a consequence of social life, yet the nature of that social life is a consequence of our being human.
>
> (Rose *et al.* 1984: 11)

But a dialectical analysis also seeks to understand change. In part, the social parameters of economic activity set limits to individuals' potentials, limits that at times frustrate people from reaching their capabilities: for instance they may become unemployed. And frustration may become anger and resentment as people realize that their fortunes are limited by forces beyond their control. For the abstract labour theory of value individuals are understood to be creative, sentient beings: they are not passively endowed with tastes and talents, nor socialized to adopt social and economic roles, but actively develop themselves through their experience (see below). Because people are creative, continually adjusting their ambitions and behaviour in the light of their achievements, it can never be fully predicted how people will react to individual frustration. And people's responses, apart from reflecting their own ambitions, expectations and potentials, will also be contingent on other individuals' reactions: the whole (society) cannot be known without understanding the parts (individuals), and the parts cannot be understood without knowing the whole.

Knowledge is a process of enlightenment. As we know more about people, and conceptualize behaviour, so we shall better understand society; and such an understanding will facilitate a deeper knowledge of people in general and themselves in particular. In studying economics, to

> understand the concepts fully requires that we understand the inner logic of capitalism itself. Since we cannot possibly have that understanding at the outset, we are forced to use the concepts without knowing precisely what they mean.
>
> (Harvey 1982: 1–2)

Human economic motivation

People, then, are not independent individuals who need to be free to realize their endowment of tastes and talents, conditioned only by their subjective preferences and the controls on personal behaviour to limit infringements on others' personal liberties. Nor does individuals' behaviour have to be managed in the common interest through the creation of an institutional environment conducive to economic cooperation. Rather, people are creative, social beings, who only express, and become aware of, their potentials *through* social interaction. And relations between individuals are in constant flux as people learn about themselves, and constantly react to others. Individuals constantly change and develop because of their social experience: a process known as *praxis* (see Chapter 17): 'all history is nothing but a continuous transformation of human nature' (Marx 1936: 124).

Economic analyses are intended to reveal why individuals relate in particular ways so as to socially cooperate to fulfil each others' economic needs, and to identify sources of frustration and the potentials for social and economic change. Economic activity has to be theorized to make explicit the essential basis of the appearance of relations of exchange, and to highlight economic contradictions: contradictions that particularly prejudice the powerless to the advantage of the powerful.

The appearance of essentially contradictory class interests:

> '[H]igh-net-worth individuals' (HNWIs) . . . 'individuals with net worth of at least a million dollars' . . . [compose] a tenth of one per cent of the world's population . . . They hold 17 trillion dollars . . . the combined GNP of two billion people who live in India and China is just over one trillion dollars . . . [The assets of HNWIs] grew by 10 per cent . . . in 1996 . . .
>
> (Foot 1998)

> Jan Leschly . . . the highest paid executive in Britain as boss of the drugs giant, SmithKline Beecham . . . earns more in an hour than a teacher in

London does in a year ... What you do and what you get paid have only the most tangential connection.

(Guardian 17 April 1998: 19)

The abstract labour theory of value attempts to explain this 'tangential connection', making the essential economic relations explicit. People, through no fault of their own, face social constraints that frustrate them from fulfilling their potentials, and such frustrations become the harbinger of social change, as the disadvantaged challenge the hegemony of the privileged in an attempt to change the balance of social forces in the control of society.

> Men make their own history, but they do not make it just as they please, they do not make it under circumstances chosen by themselves, but under conditions directly encountered.
>
> (Marx 1950: 255)

The thrust of the analysis is to identify contradictory economic interests by which the disadvantaged are constrained from living a full life. And these contradictions suggest possible political strategies for the powerless to overcome their shared frustrations through collective action. Hence the first step is to make explicit what these contradictions might be.

The mode of production

To survive, people have to transform the natural environment into useful things: sources of utility, or 'use values'. And Marx conceptualized this process in terms of the *forces of production* and the *relations of production*. The forces of production refer to the productive ability to transform the natural environment into use values: the relations of production define the social relations through which people interact to cooperate to be able to produce. But these two concepts are not empirically distinguishable.

> Productive forces and social relations are ... to be regarded as two aspects of the *same* material labour process ...
>
> (Harvey 1982: 99, emphasis added)

The mode of production is a dialectical concept, understanding the process through which people relate to each other in their material existence. And these relationships are in constant change as people facilitate and constrain each others' potentials.

> [A]s men develop their productive forces, that is, as they live, they develop certain relations with one another ... the nature of these relations is bound to change with the change and growth of these productive forces.
>
> (Marx 1975a: 34)

And in the experience of changing society people themselves change: 'By thus acting on the external world and changing it, he [humanity] at the same time changes his own [human] nature' (Marx 1974: 173).

The dialectical unity of the forces and relations of production make up the *mode of production*. And the mode of production defines the social parameters of human existence, and hence the way in which people relate to each other.

> The mode of production of material life conditions the social, political and intellectual life-process in general.
>
> (Marx 1976a: 3)

Note that the mode of production conditions, but does not determine, 'the social, political and intellectual life-process in general'. It is people's changing technical (forces of production) and social (relations of production) relations that dialectically interact to influence individuals' behaviour: and it can never be known what will be the outcome of these interactions. Events will reflect the creative social potentials of individuals, and the material conditions of existence that define the parameters of social activity.

The degree to which certain individuals control the social means of production – their ability to exercise control over other people's material existence – defines the *class structure* of the mode of production. The 'class structure' suggests contradictory interests: different rationalities to social behaviour, which conflict. And that class that holds political power is politically *hegemonic*, and will be able to impose a particular interest – a particular organization of social and economic life – on the majority.

For the abstract labour theory of value, the essence of power in society reflects the control of the social means of production, which often belies the appearance of power achieved through periodic elections in democratic societies. 'The ruling ideas of each age have ever been the ideas of its ruling class' (Marx and Engels 1952: 72); 'the class which is the ruling *material force* of society is at the same time its ruling *intellectual* force' (Marx and Engels, *The German Ideology*, quoted Reiss 1997: 75).

Economic analyses focus on economic relations between people, and hence on change in society consequent on people's changing experience as they struggle to fulfil their creative potentials in spite of social constraints. And making explicit the contradictory class interests that underlie economic relations between people is the starting point.

The capitalist mode of production

> Marx wrote *Capital* to put a weapon in the hands of workers. In it he presented a detailed analysis of the fundamental dynamics of the struggles between capitalists and the working classes.
>
> (Cleaver 1979: 3)

Power in the capitalist mode of production reflects the control of the means of production by the capitalist class – the *bourgeoisie*. And counterpoised to the capitalists are the workers – the *proletariat*.

> By bourgeoisie is meant the class of modern Capitalists, owners of the means of production and employers of wage labour, by proletariat, the class of modern wage-labourers who, having no means of production of their own, are reduced to selling their labour power in order to live.
>
> (Marx and Engels 1952: 40, footnote a)

The capitalist mode of production is by far the most technically efficient organization of production the world has ever known.

> The bourgeoisie, during its rule of scarce one hundred years [written in 1848], has created more massive and more colossal productive forces than have all the preceding generations together.
>
> (Marx and Engels 1952: 48)

Understanding the essence of the capitalist mode of production is fundamental to understanding the appearance of life in the contemporary world, and to identifying potentials for future change.

Why should 'Tomar Industries, one of several American companies that puts baseballs together in Haiti, pay its workers 38 cents for every dozen balls sewn. The average girl can sew three and a half to four dozen baseballs a day, that's $1.33 to $1.52 a day' ? (*Los Angeles Times*, 8 July 1974, cited Pilger 1998: 65). The answer on the one hand is the need of employers to maximize profits, the purpose of production within the capitalist mode of production; and on the other hand is that the workers, who do not control their means of production (machinery, raw materials, technical knowledge, marketing organization, etc.), because of the threat of unemployment and in the absence of any effective trades union organization, have to take whatever work they can get. 'Baseballs are sewn in Haiti because of desperation' (*Los Angeles Times*, 8 July 1974, cited Pilger 1998: 65).

> Haiti is the poorest country in the western hemisphere . . . half the children die before they reach the age of five . . . Most American companies pay [workers] as little as they can get away with . . . [many of which] produce goods for that symbol of all-American wholesomeness, the Walt Disney Company. Contractors making Mickey Mouse and Pocahontas pyjamas for Disney in 1996 paid eight pence an hour. The workers are all in debt, knowing that if they lose their jobs they will join those struggling against starvation. In 1990 Father Jean-Bertrand Aristide won a national election with a modest programme of reform . . . Within two years he was overthrown by generals trained in the United States and trained by the CIA . . . all nonsense about redistributing wealth [was

dropped] . . . . and . . . a World Bank Structural Adjustment Programme [accepted] . . . that would ensure that the baseballs, Mickey Mouse ears and Pocahontas pyjamas kept coming . . .

(Pilger 1998: 66)

Such injustices are not restricted to the poorest countries in the world. On the Aylesbury estate in South London, recently visited by Prime Minister Tony Blair, 59 per cent of households are so poor they are on housing benefit, 17 per cent are registered unemployed, and 78 per cent of 17-year-olds are not in full-time education (*Independent*, 3 June 1997, cited Pilger 1998: 79).

Such frustrations are not the exception in capitalist society, and the intellectual project of the abstract labour theory of value is to understand how and why the capitalist mode of production produces such economic inequalities and the resultant social degradation and disadvantage, which frustrates people from fulfilling their creative potentials, and yet still persists.

Individuals' 'private', individual economic activity becomes *social* through market exchange. Capitalist 'commodity production is ultimately social in nature; it is production not for private use but for exchange with others' (McNally 1993: 162). For Marx the essence of the capitalist mode of production is *commodity exchange*, through which free market exchange is to the advantage of a particular class interest, that of the bourgeoisie, without any conscious regulation to that effect.

> What I proceed from is the simplest social form in which the product of contemporary society manifests itself, and this is the 'commodity'.
>
> (Marx, quoted Dragsted 1976: 44)

It is fundamental to the capitalist mode of production that labour power, the ability to work, is itself a commodity. People's futures, their expectations and the fulfilment of their potentials are consequent on the vagaries of market forces: forces that are beyond *any* individuals' control. And the class advantage conferred on the bourgeoisie by commodity exchange is the profits received as a consequence of exchange: albeit the capitalist class has to compete within itself for a share of the 'surplus product' (see below). The proletariat, as a class, receive in income less than the value produced. The workers are exploited, but they are not cheated. Value reflects the abstract labour embodied in a commodity, and, in the process of commodity exchange, commodities produced by labour are valued more highly than the commodity, labour, itself. Exploitation is not a consequence of greedy and rapacious capitalists, although certainly these exist. And even if capitalists wanted to, they could not change the exploitative nature of relations of market exchange. It is not a question of capitalists exploiting their workers, but of the capitalist class unconsciously exploiting the working class. And some highly paid members of the working class, who do not control the means of production, feel relatively prosperous and certainly not exploited; but for Marx they are. It is a question of understanding the nature of social relations within the capitalist mode of production, not an issue of describing individuals' lifestyles.

> Every product of labour is, in all states of society, a use-value; but it is only at a definite historical epoch in a society's development that such a product becomes a commodity . . . when the labour . . . becomes expressed . . . as its value.
>
> (Marx 1974: 67)

Value and exchange

For qualitatively different use values to be exchanged, in terms of relative prices, there must be at least one quality in terms of which they can be valued. Different commodities have to be made commensurable, and quantitatively compared. And the one common quality of *all* commodities is that all are produced by labour. Individuals' particular skills, their concrete labour – what they actually *do*, as teachers, miners, drivers, managers, bankers, etc. – has to be *abstractly* valued, according to the social demand and social supply (not merely individuals' supply and demand) of the commodity. Individuals' 'concrete' labour is valued by the market as social 'abstract' labour.

> The equalisation of . . . different kinds of labour can be the result only of . . . reducing them to their common denominator, viz., expenditure of human labour-power or *human labour in the abstract*.
>
> (Marx 1974: 78, emphasis added)

The commodity is a social relation between the producer and the consumer. And as a concept the commodity explains individuals' social economic behaviour, which, reflecting the understanding of humans as creative beings, changes in response to other individuals' activity. In order to be able to theorize and understand changing social behaviour, the commodity is conceptualized as a dialectic.

As a relationship, the commodity, for the consumer, is a *use-value*. It satisfies a qualitatively defined want or need of individuals as consumers. For the producer, the commodity is a source of income, an *exchange-value*, quantitatively defined. Commodities are produced for exchange, not for use (by the producer).

> As the *exchangeable value* of commodities are only *social functions* of those things, and have nothing at all to do with their *natural* qualities [as use-values/utilities], we must first ask, What is the common *social substance* of all commodities? It is *Labour* . . . not only *Labour*, but *Social Labour*.
>
> (Marx 1975a: 34, emphasis in original)

It is the relation between consumers and producers that defines 'abstract labour', and the value of a commodity reflects the socially necessary abstract labour time embodied in the commodity relationship. And social necessity reflects the social need, or demand, for a commodity, and the social ability to satisfy this demand through production.

Social demand:

> For a commodity to be sold at its market value ... the total quantity of social labour used in producing the total mass of the commodity must correspond to the quantity of the *social want* ... Competition, the fluctuation of market prices which correspond to the fluctuations of demand and supply tends to reduce to this scale the total quantity of labour devoted to each kind of commodity ... If the demand for ... [a] particular kind of commodity is greater than supply, one buyer outbids another ... and so raises the price ... If conversely, the supply exceeds the demand, one begins to dispose of his goods at a cheaper rate and the others must follow ... which reduces the socially necessary labour to a new level.
>
> (Marx 1972a: 192, emphasis added)

Social supply:

> [*T*]*he quantity of labour necessary* for its production in a given state of society, under certain average conditions of production, with a given social average intensity, and average skill of labour employed ... if with modern means of production, a single spinner converts into yarn in one day, many thousand times the amount of cotton which he could have spun during the same time with a spinning wheel, it is evident that every single pound of cotton will absorb many thousand times less of spinning labour than it did before ... The value of yarn will sink accordingly ... The greater the productive powers of labour ... the smaller the value of this produce. The smaller the productive powers of labour ... the greater its value.
>
> (Marx 1975a: 38–9, emphasis in original)

The logic of social supply and the determination of value has important implications for an understanding of economic development. If an industry fails to keep pace with the general evolution of productivity, and production is still carried out with anachronistic, inefficient techniques and raw materials, then, even though the workers work as hard as they have always done, they will produce less value and consequently their incomes will fall: they will become impoverished. The value of a commodity, in the process of market exchange, is compared with the value of all other commodities against which it might be exchanged: and the price of the commodity falls. In this context, and with regard to the development of the capitalist mode of production in Africa, see Sender and Smith (1988), Chapter 5.

> [I]t is not wage-labour which determines value ... it is a question of social labour-time in general, the quantity of labour which society generally has at its disposal, and whose relative absorption by the various products determines, as it were, their respective social importance.
>
> (Marx 1972a: 882)

In as far as the prevailing forces of production mean that a use-value can be produced, supplied, with a small (large) amount of labour, the value of a commodity will be low (high). In as far as the prevailing relations of production imply that individuals desire/need, demand, a little (lot), the value of the commodity will be low (high). Value is determined only in the dialectical interaction of supply and demand. Value does not exist prior to exchange: it is not an inherent quality of commodities. Value, like the commodity, is a social relation. The subjective preferences of individual consumers dialectically relate to the technical (efficiency) and social (distribution) exigencies of the production process – the cost of production. The abstract labour theory of value is a dialectic of the subjective preference theory of value *and* the cost-of-production theory of value. They are different interpretations of the same experience. The citizen as the economic dynamic is inclusive of the consumer and the producer as conceptions of the determinant of economic activity.

The value of a commodity is revealed only when it is actually exchanged. It is akin to the idea of revealed preference in the subjective preference theory of value. The real value of a commodity is what individuals are willing to give up to acquire/consume it: the *opportunity cost*. Value is like weight: it can only be quantified by comparison with something else. Weight cannot be observed; the weight of an object is quantified by comparing it with a standard of weight, labelled 1 pound, 1 kilo, 1 ounce, etc. Similarly with value. The value of a commodity is known by what people are willing to exchange for it (the opportunity cost). So the value of a commodity is manifested only as an *exchange value*, which is not the same thing as value, but is the only way in which value can be observed. The exchange value of a commodity is expressed through money, as a *price*.

Price is not the same thing as exchange value, which is not the same as value. Money is the general equivalent through which value is expressed as an exchange value, appearing as a price. Value is the dialectical, social essence; price is the experiential, individual appearance. The latter, to be understood, has to be explained in terms of the former. Monetary theory within the abstract labour theory of value is a much neglected area of scholarship, but see: Marx (1973), Chapter on money; Weeks (1981), Chapters VI and V; and Harvey (1982), Chapters 9 and 10.

Value and crisis

Because value is a product of relations of exchange, there can be economic crises – crises that precipitate social change.

> For it is in the nature of the commodity form that the translation of use-value into exchange-value, of concrete into abstract labour may not occur . . . It is an ever present possibility that some producers will fail the test of the market, they will be unable to find a buyer for the use-values they have produced.
>
> (McNally 1993: 178)

Obviously, capitalist producers will estimate what a commodity will sell for and forecast future sales, and compare this with the estimated costs of production to predict profitability.

> The restless never-ending process of profit-making alone is what . . . [the capitalist] aims at. This boundless greed after riches, this passionate chase after exchange-value, is common to the capitalist and the miser; but while the miser is merely a capitalist gone mad, the capitalist is a rational miser.
>
> (Marx 1974: 151)

Capitalists invest in production, in the expectation of realizing a profit. But what is the source of profit, and why should producers have to compete to earn it?

Producers advance money for production: they purchase machinery, raw materials, labour power, technical knowledge, etc., and yet the product is sold for more than the cost of the inputs into the production process. Why? 'Where . . . does profit come from under conditions of fair exchange?' (Harvey 1982: 22).

The answer lies in the particular characteristics of one commodity: the commodity *labour power*, which produces more exchange value than it is worth. The exchange value of labour power reflects the commodities that have gone to produce the commodity labour power.

> A certain mass of necessaries must be consumed by a man to grow up and maintain his life . . . [and] he wants another mass of necessaries to bring a certain quota of children that are to replace him on the labour market . . . Moreover, to develop his labouring power, and acquire a given skill another amount of necessaries must be spent.
>
> (Marx 1975a: 45)

But like any other commodity, labour power also has a use-value (to the consumer, the capitalist). And the use-value of labour power is that it produces more commodities, more values. The use-value of labour power is greater than its exchange-value. Labour power produces more values (its use-value) than are consumed in the production of labour power (its exchange-value): the difference is *surplus value*. 'It must never be forgotten that the production of . . . surplus value . . . is the immediate purpose and compelling motive of capitalist production' (Marx 1972a: 243–4). It is because labour power is a commodity to be bought and sold that surplus value is produced. There has been the exchange of commodities for centuries through trade, but the defining characteristic of the capitalist mode of production is when labour power itself becomes a commodity: when commodities produce commodities. And when people's productive activity, their livelihoods and survival, their ability to work, reflect the vagaries of markets, this requires a change in society.

Labour power as a commodity is paid what it is worth, but the value produced by labour power is more than this. So, there is 'exploitation' – people do not control the product of their labours – but no cheating – people receive in income what they are worth as determined by the market, as with any commodity. That is, people are rewarded according to their exchange-value, but employed for their use-value.

> Why is the capitalist interested in buying labour power from the proletariat under conditions of 'equal exchange', i.e. at the real value of that labour power . . . Here appears Marx's main economic discovery, his *theory of surplus value*.
>
> (Mandel 1983: 190–1, emphasis in original)

Just as value is a social relation that appears as price; so surplus value is a social relation that appears as profit, rent and interest. The concepts of price and profit are not the same as value and surplus value, but are understood through these concepts. Surplus value conceptualizes the production of value over and above the reproduction of society: value that can be used for the expansion of production – what Marx calls *expanded reproduction*. This surplus is expressed as the incomes *interest, rent* and *profit*: incomes that are not a reward for the production of use-values (Marx 1972a, Parts V and VI).

Within the abstract labour theory of value, all value is produced by social labour, and hence surplus value is also a product of labour. Surplus value is a social surplus, which is distributed to individuals as interest, rent and profit: incomes which derive from property ownership, and not from production.

> Surplus value originates in the production process by virtue of the class relation between capital and labour, but it is distributed among individual capitalists according to the rules of competition.
>
> (Harvey 1982: 61)

And this distribution is a consequence of market exchange, according to the relative efficiency of competing capitalist enterprises. That is, competitive markets enforce common standards of efficiency on *all* producers in *all* sectors of the economy. '. . . competition puts a further obligation upon the capitalist to keep pace with the general process of technical change' (Harvey 1982: 88).

> [C]ommodity producers experience the law of value – external exchange governed by socially necessary labour time – as an external pressure. Should they fail to produce efficiently enough, the prices which rule the market will be insufficient to redeem their actual costs of production. The result will be a failure of self-reproduction of the producing unit [bankruptcy].
>
> (McNally 1993: 179)

However, while profit is a return based on the efficiency of the individual enterprise – the efficient use of *all* resources, means of production *and* labour power – value and surplus value are socially produced only by labour power. To remain relatively efficient and hence profitable, producers have continually to reduce the abstract labour time embodied in commodities. Enterprises have to stay one step ahead of their rivals in a competitive economy. This is achieved through technical change: by continually improving production processes, replacing living labour with machinery. By becoming more efficient, by embodying less abstract labour in commodities, the individual enterprise will become more profitable; but for the capitalist mode of production, relatively less labour is employed compared with the investment in the means of production, reducing the capacity to produce surplus value to be distributed as profit. Hence for the less efficient enterprises profits will fall as a consequence of the greater efficiency of the more profitable enterprises. 'The antagonism between each individual capitalist's interests and those of the capitalist class as a whole . . . comes to the surface' (Marx 1972a: 253).

It is the class conflict between capital and labour, for the latter to produce surplus value in the interests of the former in the context of competitive market exchange, that renders the concrete labour of individual workers abstractly social through commodity exchange. The profitability of capital in general, and the least efficient enterprises in particular, will begin to fall as the production of surplus value increasingly fails to keep pace with the need to earn profits on an ever greater investment in the means of production. That is, unless there is an increase in the rate of exploitation of labour power. Either the intensity of work could be increased – more is produced in the same time – or the amount of the working day that goes towards 'reproducing' labour power – the standard of living of the proletariat – could be reduced, compared with the amount of the working day that goes towards producing surplus value. That is, either there is a wage cut, or – through inflation – real wages fall, or the efficiency in producing the means of subsistence increases (either through labour working harder, or new technology improving productivity, but also reducing employment), so that the value of the same material standard of living reduces: hence the exchange value of the *same* labour falls.

The contradictions in capitalist production are never fully resolved: it is an ongoing process of producers having to reconcile the exigency of profitability with the terms of employment of labour power: 'just as the most fundamental contradiction of capitalism is between the classes, so the most fundamental role of crisis-as-solution is restoring the balance of class forces such that capital can resume its growth' (Bell and Cleaver 1982: 257).

For Marx this is not an anomaly but a contradiction. For individual enterprises experiencing falling profits, the *only* response can be either to directly exploit the labour force more effectively, which will create tension and struggle, and/or to invest in a new production process, whereby the same workforce works with more technically sophisticated means of production. This latter option may make the enterprise more profitable, but this is at the cost of reducing the ability of the capitalist mode of production to generate the surplus value that is manifested as profits, *without* increased exploitation. And an augmentation of exploitation tends to beget tension and struggle.

'The *real barrier* to capitalist production is *capital itself*' (Marx 1972a: 250, emphasis in original).

> [W]hat is rational *from the standpoint of the system as a whole* is not rational from the standpoint of each ... firm taken separately, and vice versa ... The naive conviction that the 'common interest' is perfectly served if each individual pursues his 'private interest' turns out to be manifestly illusory ...
>
> (Mandel 1978: 178–9, emphasis in original)

The particular ways in which the contradictions of capitalism are experienced – as unemployment, wage cuts, inflation, worsening conditions of employment, bankruptcy, interest rate rises, reduced state expenditure, privatization, etc. – will be historically specific, reflecting the creative potentials of different people, at different times and in different places, and changing material conditions of their social existence. And these experiences will also be contingent on the degree to which people are aware of their class interests: their ability to rationalize their struggles over wage levels, employment, education, health provision, discrimination, and inequality as class issues. But being class-conscious is only half the story. People also have to be able to mobilize and organize themselves as a political force struggling for change.

> The difficulty is that there is not one Marxist theory of economic crisis, but several ... all Marxist perspectives on economic crisis tend to see crisis as growing out of the contradictions inherent in the process of capital accumulation ...
>
> (Wright 1975: 5)

And the process of capital accumulation is the generation of surplus value and its distribution as profit. This whole analysis is generally referred to as the *tendency for the rate of profit to fall* (Castells 1980, Chapter 1; and Cole 1995, Chapter 5).

> The rate of profit does not fall because labour becomes less productive, but because it becomes *more* productive . . . The rate of profit does not sink because the labourer is exploited any less, but because generally *less labour is employed* in proportion to the employed capital.
>
> (Marx 1972a: 240, 246, emphasis added)

This contradiction is of the capitalist mode of production, a mode of production that is ever more global 'if there is a secular tendency of . . . the rate of profit to decrease, this can only be proved by studying, in terms of value [and hence in terms of "essence" rather than "appearance"], the process of accumulation at a *world level*' (Castells 1980: 40, emphasis added). On the oscillation of economies between periods of growth and recession see Mandel (1977), Fischer (1996), and Van Duijn (1983).

> The bourgeoisie cannot exist without constantly revolutionising the instruments of production, and thereby the relations of production, and with them the whole relations of society . . . The need for a constantly expanding market for its products chases the bourgeoisie over the whole surface of the globe. It must nestle everywhere, settle everywhere, establish connexions everywhere.
>
> (Marx and Engels 1952: 45, 46)

And the 'constant revolutionizing of the instruments of production' at a world level had led to a third of the world's workforce, over 1 billion workers, being underemployed or unemployed by September 1998 (report from the International Labour Office, 23 September 1998). According to Michel Hansenne, director general of the Geneva-based organization, 'The global employment situation is grim and getting grimmer' (reported in *Guardian*, 24 September 1998: 19)

Economic experience and economic analysis: the East Asian economic crisis and beyond

The tendency for the rate of profit to fall is an essential process, defined in terms of value, unlike the appearance of economic activity, expressed in price (financial) terms, and this tendency may be offset by counteracting tendencies, for instance an increased rate of exploitation, and thus may not be directly empirically observable. However, this ever-present tendency explains the trajectory of economic policy, intended to maintain the rate of profit and the process of capital accumulation in spite of this trend.

Capital accumulation may not be an explicit policy objective but, as we have seen, competitive forces in market exchange socially value as abstract labour the work and concrete labour of individuals, quite independently of any policy prescriptions. And although free market exchange to maximize the exploitation of wage labour in the production of surplus value is in the class interest of capitalists, it is in the interests of *individual* capitalists to exert monopoly power to bias the relations of exchange to serve 'their' share of surplus value maximizing 'their' profits – their share of surplus value. And the process of economic crisis in the late twentieth century is all about the guardians of the class interest of international capital, essentially the IMF (International Monetary Fund), World Bank, and the GATT (General Agreement on Tariffs and Trade; now superseded by the World Trade Organization, WTO), prevailing upon recalcitrant individual capitalists to prosecute the best interests of the world capitalist class *vis-à-vis* the world proletariat.

To illustrate the intellectual complexity of trying to interpret the appearance of economic activity through a conception of the essence of experience, on a *world* level, this final section will address the current process of economic crisis (that is, current at the time of writing, October 1998), highlighting the significance of a class analysis.

In 1998 the world economy has been reacting to the collapse of the Far Eastern economies: a major turning point in the world economy, which as we shall see has had a whole raft of implications for countries around the world, but which was a complete surprise to orthodox economists responsible for the smooth running of international trade and finance. A 'seemingly isolated financial bust that started last summer [1997] in Bangkok has escalated into the biggest threat to global prosperity since the oil shocks of the 1970s' (Bremner 1998). Based on forecasts of continuing economic growth and rising prosperity the international monetary authorities were extremely sanguine.

> East Asia's capacity to sustain rapid growth is *without precedent* . . . East Asia enjoys *enviable* macroeconomic conditions and an unprecedented growth momentum.   (World Bank 1997: 1–2, emphasis added)

> Most economic specialists predict that this year's growth rates of East Asian economies will be higher again than expected.
> (Paul Samuelson, Nobel laureate economist, quoted Gyoung-Hee 1998: 39)

> The fundamentals of Korea's economy themselves are sound.
> (Michel Camdessus, managing director of the International Monetary Fund, May 1997, quoted Gyoung-Hee 1998: 40)

Yet less than a year later the 'miracle economies' of South-East Asia (Thailand, Malaysia, Indonesia, Philippines) and East Asia (Korea, Taiwan, China), and latterly Japan, were in crisis. It is misleading to see this as a particularly East Asian phenomenon: it is a moment in the ongoing process of the world conflict between contradictory class interests reacting to the dynamic of the accumulation of capital. The East Asian crisis is only the latest episode in the ongoing saga of capitalist crisis, in the world, capitalist mode of production. In 1992/3 the problems were in Europe; in 1994/5 Latin America; 1997/8 East Asia; in November 1997 *The Economist* warned that 'Brazil is near the edge . . . Its congressmen should help the government to save it from going over' (*The Economist* 22 November 1997). Brazil is important in that it accounts for 45 per cent of the total gross domestic product of Latin America.

And Will Hutton, William Keegan and Ed Vullamy warned in the *Observer* (16 November 1997) that Argentina would follow Brazil: 'if Brazil falls into recession . . . it will bring down Argentina.'

Returning to East Asia, the countries in this region have significantly different economic histories and potentials. Japan was a substantial industrial power by 1970; in the 1970s Singapore, Taiwan and South Korea experienced rapid economic growth, followed in the 1980s and 1990s by China, Malaysia, Thailand, Indonesia and the Philippines.

> To listen to the recent debate about the death of the Asian economic model, you would think that all the tigers shared the same policies and problems. In reality, though, there are huge differences between them. For example, the role of the state ranges from hands-off in Hong Kong to heavily interventionist in South Korea, Indonesia and Malaysia . . . and all manner of things in-between elsewhere.'
>
> (*The Economist*, 7 March 1998: 14)

That there was rapid and remarkable economic growth in the region in part reflected that these were essentially agrarian societies that were being transformed into industrial powers. Expanding modern industry could draw upon cheap peasant labour from the countryside, which was rapidly proletarianized (dispossessed of the means of production), becoming (landless) *wage labour*.

> Born into a world of agriculture, they live in a world of industry. Born in villages they live in cities. Born to obey the old ways, they struggle in confrontation with the new.
>
> (Sparks 1998: 11)

Such rapid economic growth and social change is very disruptive: some parts of society are fundamentally transformed while others are scarcely touched. A large, cheap labour force, struggling to escape widespread, 'precapitalist' rural poverty is not an East Asian phenomenon. This was exactly the process by which capitalism emerged in Europe in the eighteenth and nineteenth centuries (e.g. Engels 1973). And there has been considerable state repression in East Asia to keep this labour force 'cheap', to facilitate the emergence of a local, industrial, capitalist ruling class. Opposition to, and oppression of, any form of organized labour with socialist pretensions has been common throughout the region, although such denial of human rights was at its most extreme in Indonesia, when in 1965 President Thojib Suharto led a coup that 'unleashed one of the worst massacres the world has seen . . . Hundreds of thousands . . . were slaughtered, the names of many supplied by the CIA. Virtually every radical element in the country was eliminated' (Fermont 1998: 13). The final death toll is estimated at up to 1 million. But though particularly brutal, Indonesia was not the exception (Sparks 1998: 14–15).

> Formal ideologies aside, one thing the rulers of the region have in common is ruthless hostility to working class organisation and struggle. But that, as we know only too well, is not an especially 'Asian' characteristic.
>
> (Sparks 1998: 15)

There has been a close relationship between the state, banks and industry in the region. Local capitalists needed state support to compete against technically more advanced, foreign rivals, through such measures as tariff protection, incentives to exporters, and cheap credit. A rift began to appear between the interests of local capital and the operation of the world market as a process of class exploitation in the accumulation of capital. Expansion was fuelled by easy borrowing: 'generous loans [were made] to local industrialists at advantageous rates' (Sparks 1998: 18).

Because the benefactors of cheap loans were often family and friends of those in power, the economies and economics of the area are often called *crony capitalism*. Referring to the collapse of the South Korean economy:

> The *chaebol*, the crony-run conglomerates that dominate the economy, having been allowed to borrow too much, have expanded without regard for adequate returns . . . their profits are lousy.
>
> (*The Economist* 29 November 1997: 25)

While it is the 'restless, never ending process of profit making' (Marx) that is the dynamic of competitive capitalism, individual capitalists will try to steal a march on their rivals in their acquisition of surplus value as profits. And in South Korea enterprises were insulated from the full force of competitive pressures.

The dynamic of the (world) capitalist mode of production, the competitive law of value, by which value reflects the 'socially necessary abstract labour time' embodied in commodities, could be denied only by ever greater, disguised subsidies – cheap loans.

Yamaichi, one of the world's top 10 investment houses, hid more than
£1 billion in suspect loans made to preferred customers through the off-
shore financial haven of Cayman – a practice known as *tobashi* in Japan.

(*Guardian* 24 January 1997)

Eventually the chickens came home to roost. With the East Asian economies
accounting for nearly 45 per cent of Japan's exports, the chilling effects on the
Japanese economy of the cold economic winds blowing through East Asia were
inevitable. Banks and finance houses collapsed under the weight of bad debts.
On 13 January 1998 the *Guardian* reported that Japanese banks held some
76.7 trillion yen in loans (£348 billion), 2.7 trillion of which was judged to be
irrecoverable. 'Swathes of Japanese companies are insolvent in all but name,
many of them are going to go under' (*The Economist* 11 April 1998: 17). In the
first quarter of 1998 corporate profits fell by over 40 per cent and at the end
of September the Japanese Leasing Corporation (JLC) filed for bankruptcy,
marking the biggest financial failing in the history of Japan, involving 2.2 trillion yen (£10 billion) of debt. The JLC, a subsidiary of the Long Term Credit
Bank of Japan (LTCB), threatened the whole banking sector, and the LTCB
had to be nationalized to contain the threat. Already by the end of May Japan's
eighteen largest banks had written off £45 billion in bad debts. By September
the Japanese economy carried more debt that the entire Canadian GDP.

Such developments were repeated elsewhere in Asia. In Thailand 'bad debts
soared to equal a quarter of banking assets and the stock exchange fell 82%'
(Harris 1998: 23). And in August 1998 some 2000 people a day were losing
their jobs, with a jobless total of around 2.2 million, unprecedented for
Thailand (*The Economist* 8 August: 76).

To address the financial chaos, economies unable to maintain the value of
their currency in the face of intense international speculation have had to
turn to the IMF for support, which has invested the International Monetary
Fund with the power to enforce capitalist, competitive rules in the interests
of the world capitalist class as a whole. The law of value is unwittingly
enforced. As a caption to a picture headed 'The moment of surrender:
Indonesia's dictator bows to the man from the IMF' (*Observer* 10 June 1998)
underscores: 'Michel Camdessus, head of the International Monetary Fund,
watches President Suharto accept his terms in Jakarta on 15th January'.

[In November 1997] . . . the IMF was in a position to prise open the
Korean market, to end government directions to the banks on who they
should lend to, to shut down the insolvent banks, to allow in on terms
of equality, foreign capital to restructure the *chaebol.*

(Harris 1998: 23)

Value, under the capitalist mode of production, is defined only in the
process of competitive exchange. Any political interference in this process,
whether by governments or by those with monopoly economic power, can

be maintained only by ever greater subsidies in one form or another, or intensified exploitation: in this case competitive exchange was mollified through 'cheap' loans. And the IMF is merely the debt collector for the capitalist class: debts that are collected through currency devaluation, interest rate rises, tight monetary control to ward off inflation, reform of the banking and financial system, the freedom of transnational capital to buy up national assets, the dismantling of protection to allow international competition, etc. And while all these policies adversely affect local capital, the cost is ultimately paid by the working class, in terms of unemployment, higher prices, wage reductions, cuts in state welfare expenditure, etc.

This is the reality, the appearance, of the essence of capital accumulation and the tendency for the rate of profit to fall. On the collapse of Japan's fourth largest stockbroker, Sanyo Securities and the Hokkaido Takushoku Bank, *The Economist* crowed, ' "Better late than never" . . . Japan's financial authorities, along with the firms themselves, are at last accepting *reality* . . .' (*The Economist* 29 November 1997, emphasis added). But it was also a reality that Japan's ruling Liberal Democratic Party was looking to support from key business interests, using taxpayers' money to support Japan's ailing banks.

On the other hand, the IMF has acted to achieve the dominance of the interests of the capitalist class over the particular interests of local capital, where the latter has turned to the state for protection: 'allowing bankrupt-cies requires political courage by governments, facing angry friends and relatives, the unemployed and voters, and the willingness of business groups to sacrifice their political size to profitability' (Harris 1998: 24). And in this regard the Indonesian state has been at great pains to protect the particular interests of President Suharto's family.

But, in the last analysis, it is the working class that will pick up the tab. Capitalist discipline is enforced whatever the human cost.

> In the 3 economies hit hardest by the region's economic turmoil – Indonesia, South Korea and Thailand – jobless rates have roughly doubled . . . Moreover, large-scale redundancies as a result of economic crisis *have barely started*.
>
> (*The Economist* 25 April 1998: 101, emphasis added)

> According to investment funds and business consultants in the region, Indonesia will shed two million jobs this year, Korea about 1.5 million, Thailand over one million, about 150,000 in Malaysia, close to 100,000 in the Philippines, and probably 20,000 in Singapore.
>
> (Report in the *Hindu* 28 January 1998, quoted Sparks 1998: 30)

> East Asia does not face crisis, it faces catastrophe. Tens of millions will be made destitute . . . The figures are like casualty lists from the First World War . . .
>
> (*Observer* 26 April 1998: 28)

Fundamentally the concrete labour worked in these economies has to be reduced to socially necessary abstract labour (the law of value), by standards set by the competitive pressures of the world market.

> Soon, and for the first time, Japan will have a higher jobless rate than America ... Redundancies and bankruptcies are evidence that the economic adjustment mechanism is working at last. *Capitalism*, you might say, *is finally coming to Japan*. Pity it had to be the hard way.
>
> (*The Economist* 20 June 1998: 23, 25, emphasis added)

Although George Soros, the international currency speculator, is not so confident that a dose of capitalist discipline is what Japan needs: 'The global capitalist system that has been responsible for our remarkable prosperity is *coming apart at the seams* ... the pain at the periphery has become so intense that individual countries have begun to opt out of the capitalist system' (quoted *Guardian* 16 September 1998: 2).

In the second week of August 1998 the world's stock markets were in turmoil. The Japanese yen plunged to an eight-year low, and in response markets in Europe lost up to 4 per cent of their value; Hong Kong closed at a five-year low; Malaysia at a ten-year low; in Russia shares were down 9 per cent, affecting the Hungarian and Czech stock markets; the collapse of commodity prices seriously weakened the New Zealand, Australian, Norwegian and Canadian economies; at the end of August the Wall Street stock market registered the second biggest fall ever, with $500 billion knocked off the indexes; Hong Kong, Frankfurt and London joined the retreat; in the first seven months of 1998 there was a 35 per cent fall in the Brazilian stock market, and to stem the outflow of finance interest rates had to rise to 50 per cent; in Venezuela the stock market tumbled by 65 per cent; Mexico had to bail the banking system by the equivalent of $700 for every man, woman and child, in a country where 40 per cent of the population live on $1 a day; the South African currency devalued by 15 per cent; the Colombian and Ecuadorian currency devalued; in the first six months of 1998 the Chilean stock market fell by 40 per cent; by September a third of the economies of the world were officially in recession; at the end of September there was an unprecedented $3.5 billion bail-out by a consortium of fifteen international banks, arranged within 48 hours by the Federal Reserve of the United States, of the Long-Term Capital Management fund, which collapsed owing at least $200 billion (after using a $4 billion capital base to wage bets on the financial markets of the world totalling $1.25 trillion), a 'hedge' fund that ironically was intended to make international private investment *less* risky; in its World Economic Outlook the IMF lowered the forecast for growth in the world economy from 4.5 per cent to 2 per cent; 14 October saw the Bank of America, the biggest banking group in the USA, announce a provision of $1.4 billion for credit losses in the third quarter on 1998; it was reported that the number of Americans filing for personal bankruptcy in 1997 rose to 1.4 million, three times the figure for 1980; etc.

Such are the machinations of international competition, which impose the abstract labour theory of value in the (world) capitalist mode of production, leading to economic crises that Marx believed 'would get worse and worse' (Bygraves 1998: 25).

Ultimately the burden is borne by labour. And as market economies fail to provide work, as the casualty lists rise, the free enterprise system begins to lose legitimacy, especially in the face of extreme economic inequality. While the rich see their assets depreciate by 'a few noughts . . . the poor . . . face having to knock a mouth or two off their families . . . While the mass of people suffer, those in power do not' (Sweeney 1998b). And as people become angry, and struggle to fulfil their potentials, there will be state repression. Class contradictions can only be overcome by challenging class power, even if the dispossessed and disadvantaged are not conscious of the nature of their struggle. At some time class issues will come to the fore, and will elicit the heavy hand of the state to protect the privilege of the local ruling class: it is left to the power and influence of the IMF to prosecute the interests of international capital as a consequence of protecting the interests of the most advanced and efficient capitalists in the world – the transnational corporations who owe no allegiance to *any* nation state.

As the 'class struggle' unfolded, mass strikes erupted in South Korea in January 1997 as 'workers . . . resisted mass sackings' (Gyoung-Hee 1998: 45); isolated incidents in Japan, strikes in Thailand, and signs of discontent in China (Sparks 1998: 33); in Indonesia 'the assumption among both businessmen and economists is that economic collapse will provoke such severe social unrest that the army . . . will be forced to decide where it stands' (*Financial Times*, quoted Fermont 1998: 12); 'Military shows mailed fist' (headline *Guardian* 6 May 1998: 13). In Malaysia, a policy of repatriating foreign labour in response to growing unemployment has led to 'some of those who resisted repatriation . . . [being] tortured and shot' (Sweeney 1998a), and, in an apparent attempt to make room in the detention camps, detainees' food has been poisoned, and there is 'evidence that the . . . poisoning was . . . part of a nationwide strategy' (Sweeney 1998a). In August there was widespread social unrest as a consequence of price rises in Indonesia, unrest that reignited in early November with tens of thousands rioting in Jakarta, leading to dozens of deaths at the hands of the security forces as unemployment grew to over 10 million, with over 60 million people living below the poverty line – on the unfolding conflict in Indonesia see Fermont (1998). In Northern Java there were six consecutive days of rioting in which rice mills, shops and warehouses were looted as a consequence of rising prices such that the daily wage could not buy even a kilo of rice. In Hong Kong, with falling prices, wage cuts, and falling property prices, the stock market had to be supported by an injection of £9 billion to stabilize share prices.

In August 1998 the process of crisis moved on to Russia. 'Raw fear was tangible in Moscow yesterday as panic rose up and submerged the International Monetary Fund's multi-billion dollar [$22 billion] wall of money that was designed to shore up confidence in Russia' (*Guardian* 14 August 1998: 20) – money which was largely spirited out of Russia into foreign bank accounts within a few weeks, and although the IMF rescue package was specifically designed to avoid devaluing the rouble to try to inject stability into the world finance markets, within a month the currency was devalued.

According to currency speculator George Soros the crisis in Russian financial markets was 'terminal', the climax of 'seven years of pro-Western rule in Russia' (Meek 1998: 18). Whereas the East Asian economies transgressed the law of value with cronyism, in Russia it was 'terminal kleptocracy' – or what Wolfgang Schüssel, Austrian foreign minister, referred to as 'hyena capitalism'. And *The Economist* (5 September 1998) noted that weak and corrupt government meant that Russia had no 'rule of law', and as a consequence state assets had been acquired cheaply through rigged 'privatizations'. The Russian debacle is not important in itself. Unlike Japan (the second biggest economy), Russia accounts for only about 2 per cent of world output (about the same size as Portugal), but by mid-September 1998 British, German, US and Japanese banks had lost about $9 billion in loans to Russia.

While it had always been conceded that conversion from the Soviet command economy to the Russian market economy would be painful for the majority of Russians, people were told that, with hard work and honesty, they would survive the transition, and that they and their children could look forward to a brighter, capitalist future. Meanwhile, those who had been at the top of the Soviet regime, the 'nomenclatura', exploited capitalist freedoms, and in the absence of an effective legal framework fraudulently acquired privatized state assets and siphoned off billions of the dollars injected by the IMF into the ailing Russian economy. By August 1998 $3.75 billion left the Russian economy for Swiss banks *every month*. With such a haemorrhage the healing balm of the IMF was ineffective. And monetary policy to restrict the money supply to head off impending hyperinflation, in the absence of IMF funds, only meant that the state had no money, and so state employees, including the army and miners, were not paid for months.

In the face of such blatant corruption the Group of Seven industrialized countries (USA, Germany, France, Britain, Italy, Japan, Canada) at the end of August refused to add to the $22.6 billion already lent to Russia until there was 'a sound basis for public revenues . . . and action to restore stability to the financial system' (*Guardian* 28 August 1998). And at the end of September there was a new twist to the tale. In the face of rising prices, rising unemployment, and millions of workers and soldiers being unpaid for months, the Central Bank announced a fundamental shift in economic policy from the liberalized freedoms promoted by the IMF and the Group of Seven, which included a tight monetary policy intended to stifle nascent inflation. Instead the printing presses were to be set free. Billions of roubles were to be printed to pay state employees and the army before revolution unfolded. Of course, the ensuing inflation will slash the value of this money almost before the printing ink is dry; and to try to impose some financial order, the banks are to be renationalized, which will lead to many privatized industries returning to 'public' ownership.

This sea change in economic policy follows the rise in political influence of the Communists after President Boris Yeltsin's failure to get his nominee for prime minister, Victor Chernomyrdin, accepted by the Duma, having to accept Yevgeny Primakov, who placed one of the planners of the former Soviet economy as head of the Central Bank at the end of September.

And so the process continues . . .

The long-term decline in the 'real' economy, as the value of commodities produced abstractly declines, and capital accumulation is compromised as the production of surplus value and the tendency for the rate of profit to fall fails to be offset by a sufficiently increased rate of exploitation, leads to falling profits, and the crisis appears in the sphere of *finance*. However, financial markets fluctuate around this declining trend, periodically giving the appearance of recovery. After the post-war economic boom, profits began to decline sometime in the late 1960s (Mandel 1977), and while there have continued to be booms and slumps, the periods of slump have become relatively deeper, more prolonged, and more frequent: hence the trend to abstractly devalue concrete labour is manifest in economic trends, and appears as a question of the declining confidence of investors – precisely Keynes's assessment in the 1930s. For instance, in October/November 1998 economic problems in Brazil appeared to be coming to a head because of hesitancy on the part of investors. When in July/August the IMF injected over \$25 billion into the Russian economy to support the rouble, and the money disappeared, the government defaulted on international debt, and the rouble was devalued, investors got cold feet (or was it cold pockets?). Money moved out of emerging markets, and Brazil, with a large fiscal deficit, the burden of which was seen as the harbinger of a potential devaluation of the 'real', was judged to be risky for speculation. In the three months from August \$30 billion fled the economy, leaving at the end of October only \$42 billion in reserves. And just as financial markets appeared to be stabilizing it was essential to try to protect the value of the real. Confidence had to be given a fillip to avert a Russian-style capital flight. Consequently on 13 November 1998 Brazil agreed to a package of austerity measures – which included budget cuts of \$23.5 billion dollars (3 per cent of gross domestic product), tax increases, reform of the pensions system, changes in contracts to make it easier to make state employees redundant, fiscal reforms to make regional government finances more independent – in exchange for a \$41.5 billion (£25 billion) rescue package (the fifth major IMF rescue in 15 months). The package requires that the \$47 billion budget deficit become a surplus of 2.6 per cent of gross domestic product by the year 2000.

The approach is all about building confidence through appropriate economic management by the fiscal authorities. For orthodox economic theorists, such crises are not a consequence of irreconcilable class contradictions. It is a problem of the state serving capitalist (class) interests: 'the vulnerability to financial crisis was created not by international speculators and other bogeymen but by woefully inadequate oversight by domestic finance, private, public and quasi-public' (*The Economist*, 5 September 1998: 13). And at the end of September 1998, in a speech to the New York Stock Exchange, British Prime Minister Tony Blair outlined a suggested regulatory framework for the world economy that would combine the IMF, the World Bank, and the Bank for International Settlements into a system that would have: greater transparency to strengthen the incentive for governments to pursue sound policies; improved financial regulation and the effective coordination of internationally active financial institutions; better risk assessment and surveillance with closer contact between public and private sectors; greater openness and accountability of international financial institutions; etc.

While the speech illustrated a move away from the 1980s and 1990s free-market, liberal orthodoxy, with the need for political authorities to address people's jobs rather than rising prices, implying a need to manage economies, the speech showed no awareness of the causes of economic crisis. The concern is over *how* crises occur, not *why* (on the methodological implications of 'how' and 'why' questions see Part IV).

For the abstract labour theory of value, the contradictory social relations of exploitation that underlie the capitalist mode of production produce a tendency for the rate of profit to fall as less labour is employed (a diminution of variable capital) with an ever greater stock of constant capital. This essential contradiction will eventually appear as falling profits, leading to firms' being unable to service their debts, and hence bank failures, higher interest rates, falling commodity and share prices, currency devaluations, rising state deficits and inflationary pressures, etc. The crisis will first appear in the financial sector. And in this regard *The Economist* (3 October 1998: 19) correctly noted that the experience of the banking sector 'consists of one speculative bubble after another ... fuelled by an explosion of credit, a wave of unwarranted optimism and a subsequent mispricing of risk.' In the process of the changing social relations of exploitation by which enterprises default on their debts, it appears as if the problem is one of incompetent bankers, ill-regulated economies and possibly corrupt governments, with the whole mess being exacerbated by self-interested speculators; that the risk has increased through class struggle, the changing social relations of exploitation, is not apparent. And consequently, it is not simply a question of working 'with others to restore confidence and manage change and to stabilize the financial system' to avoid a 1930s-style recession in the face of 'the most serious financial challenge in 50 years' (President Bill Clinton addressing the annual meeting of financial ministers and central bank governors in Washington, 2 October 1998).

Appearances are deceptive.

> The dollar was strong; the dollar is now weak. Stock markets were at an all time high; now they are in headlong retreat. Emerging markets offered long term rapid growth; now they are mired in recession. Long bonds were a safe haven; now even they lack a sufficient comfort factor. Financial markets seem to have entered a looking glass world in which all the assumptions of recent years have been turned upside down.
>
> (Phillip Coggan, 'Curiouser and curiouser',
> *Financial Times*, 10 October 1998)

It is a question of being 'class conscious' and supporting and struggling for those policies in your class interest. And the ensuing class conflict and social change will increasingly put 'socialist' development objectives on the political agenda (see Chapter 13).

> In history up to the present ... individuals have ... become more and more enslaved under a power alien to them ... a power which has become more and more enormous and, in the last instance turns out to be the *world market*.
>
> (Marx and Engels 1970: 55, emphasis in original)

Part II
Environment

Environmental theory

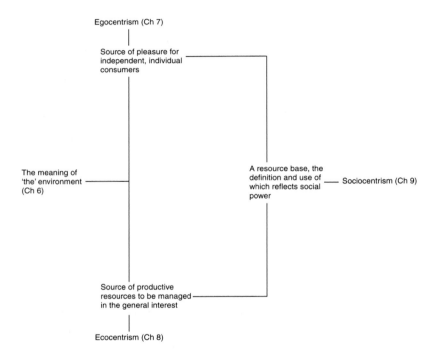

6 Environment

The environmental issue

The environment as a social, economic, and political issue has become ever more pervasive in the past twenty years or so. Governments have adopted dozens of environmental and resource policies, and new institutions have been established to monitor and manage these initiatives. In the United States, the 1970 Clean Air Act established air quality control standards with regard to such pollutants as sulphur dioxide and ozone. Individual states were given the responsibility to attain these targets under the overall jurisdiction of the Environmental Protection Agency (on US environmental policy see Tietenberg 1992 and Portney 1990). And a United States senator, who became US vice-president, published a book on environmental policy in 1992 (Gore 1992). In the United Kingdom, a leading environmental economist, David Pearce, was appointed Special Advisor to the Secretary of State for the Environment, and within a year *Blueprint for a Green Economy* (Pearce *et al.* 1989) was published, outlining a strategy for sustainable development, such that 'future generations should be compensated for reductions in the endowments of resources brought about by the actions of present generations' (Pearce *et al.* 1989: 3; see also Pearce 1993a, 1993b, 1995). The British government produced a White Paper on environmental policy in 1990 (Department of the Environment 1990), and the higher profile of environmental policy increased spending on environmental issues from £4.8 billion in 1986/7 to £14 billion in 1990/1 (Department of the Environment 1992: 237–47).

In recent times environmental issues have taken on an international importance, with the growing recognition that such problems as global warming, climate change and biogenetic hazards can only be addressed at a global level. And 1987 saw the publication of the report of the World Commission on Environment and Development, *Our Common Future*. In 1991 the Global Environment Facility (GEF) was established by the World Bank, the United Nations Development Programme and the United Nations Environment Programme to provide help to developing countries to address issues such as biodiversity, water pollution and climate change. The GEF was relaunched in June 1992 at the largest international conference ever organized, the

United Nations Conference on Environment and Development (the Earth Summit). Representatives from 179 nations (including over 100 heads of state or government), and numerous non-governmental organizations (NGOs) from around the world met in Rio de Janeiro.

The policy decisions and treaties resulting from the Earth Summit were weaker than many hoped for, and even some of these the US government failed to approve; and in 1995, at the World Climate Conference in Berlin, a follow-up meeting to the Convention on Climate Change agreed at Rio de Janeiro in 1992 to discuss greenhouse gases and global warming, and restrict the emission of carbon dioxide, the Middle East oil-exporting states wanted no commitment to limiting the burning of fossil fuels, and the United States, Canada and Australia reported that they would be unable to honour their commitment at the Earth Summit to restrict emissions in the year 2000 to 1990 levels. However, thirty small island states threatened by rising sea levels called for a 20 per cent cut in emissions from 1990 to 2005. Other than agreeing that further cuts were needed after 2000, the Conference agreed to meet again in 1997, a meeting held in Kyoto, Japan, to decide admissible levels of pollution (on this meeting see the section 'The commoditization of the environment' in Chapter 7).

But the importance of the Earth Summit, and other attempts at the definition of policy initiatives coordinated at a global level, is not so much in the detail of this or that policy, but that the global environment is on the political agenda. A political process has been initiated: as with *any* political process, vested interests will attempt to preserve their privileges, and the disadvantaged have to mobilize and organize to prosecute their interests. In this context, the various protocols emanating from the United Nations Conference of Environment and Development have been vigorously resisted by established economic and bureaucratic interests. The way forward is conceived in different contexts.

The politics of the environment

Perhaps success in policy making and implementation is to be sought in a greater understanding by the public as to the danger presented to individuals' lifestyles by our present way of life. There will be an acceptance of the need to change our behaviour in response to appropriate individual incentives. Alongside education for a new way of life, there need to be reinforced property rights over environmental resources so that individuals have a direct interest in preserving the value of 'their' assets (see Chapter 7).

Alternatively the emphasis could be placed upon reforming pluralist, policy-making institutions to address a new political reality (Sand 1995, Esty 1994). While generally there has been a decline in support for traditional political parties, support for new environmental social movements has soared (in the case of the UK, see Department of the Environment 1992: 273). And the austerity of the 1980s and 1990s has spawned a whole spectrum of local organizations and movements fighting particular issues with regard to resource allocation on such issues as education, welfare, health services, and provision for the old.

The economic recovery in the developed market economies of the mid-1990s has done remarkably little either for less developed economies, or for the poor and disadvantaged of the developed world. Economic growth and distribution have benefited corporations and high earners, while issues such as universal health provision, child care, better schools, employment opportunities for the poor, fair trade, public transport, debt relief, and other initiatives targeted at the disadvantaged have not received similar priority. Indeed, in general, there has been a reduction in provision for social services and help for the poor. The underclass has grown poorer, while the pay, benefits and share options of the rich have exploded.

> When the state fails to deliver public goods, insurance, management of externalities, minimum basic needs and democratic rights, civil organizations may fill the vacuum. The same holds for the market where market failures lead to the emergence of institutions, many of which take the form of [community] organizations.
>
> (de Janvry *et al.* 1991: 4)

This new institutional context provides the basis for the rational management of the environmental resource base in the common interest – an interest defined by pluralist politics (see Chapter 8).

There is a third option: not the reassertion of the rights of the individual, nor the achievement of an institutional compromise for the common good between conflicting interests, but a recognition that the power of privileged individuals to protect their interests lies outside political institutions, and that, as concessions are made through political channels to reallocate resources to social needs, the room for manoeuvre for further change, while maintaining the political status quo, is ever more restricted. And such a trend of increasing inequality and degradation is the harbinger of future conflict. But this anger and frustration can be channelled towards constructive political, economic and social change, which potentially will transform the way people socially relate to the environment. Contradictory interests have to be recognized, and people have to decide which 'side' they are on (see Chapter 9).

> The world's resources, environments and peoples are locked into a globally integrated system of exploitation, whose defenders and representatives, gathered together at the Earth Summit ... The struggle to rescue the planet's environment from destruction and set it on a path of genuinely sustainable, rational development is inseparable from the battle between contending classes for the organization of society in the interests of either profit or need ...
>
> (Treece 1992: 66)

People need not be passive victims of either environmental or economic forces. Social existence is not beyond people's control.

> Citizens can, if they so choose, assert that their mutual obligations extend beyond economic usefulness to one another, and act accordingly to one another.
>
> (Reich 1998: 27)

Each of these options for progress implies an image of the environment, and each will be detailed in Chapters 7–9.

Questioning the environment

There are essentially three questions to ask of the environment:

1   *What* is the environment?
2   *How* is an ecological balance maintained?
3   *Why* are there forces tending towards ecological destabilisation?

On the methodology implied by 'What?', 'How?' and 'Why?' questions see Part IV.

The first question is one of existence, highlighting the nature of human wants and needs, which defines what is known about the environment, and hence what the 'environment' *is*. For example, until oil prices rose significantly as a consequence of the monopoly power of the Organization of Petroleum Exporting Countries (OPEC) in the 1970s, it was not economically viable to explore for oil off the North Sea coast of Britain, and effectively before this time North Sea oil did not exist.

The second question is really one of the 'physiology' of the environment. How, in the face of all the internal and external forces that are acting to change the environment, is an ecological balance maintained that is conducive to human life? And what are the consequences for climate conservation, production and health?

> ... the persistence of ecological systems must be sought not in the absence of change but in the restoring processes that keep the changes within bounds most of the time, systems persist not in spite of change but *because* of the changes they experience.
>
> (Haila and Levins 1992: 58, emphasis added)

The third question is an historical one. What is the pattern of social evolution, and how does the change in people's behaviour impact on the environment? In this context, Yrjö Haila and Richard Levins suggest three fundamental reasons why social activity despoils the environment: 'greed, poverty and ignorance' (Haila and Levins 1992: 60).

- Greed: the imperative in a capitalist world economy is for enterprises to be profitable. The necessity of putting short-term financial gain before long-term social interests: profits before need.
- Poverty: a transition towards economic activity directed at restoring nature requires reserves. Where there is desperate poverty, political authorities are rightly more concerned about food supply and employment than about the harmful effects of pesticides and industrial development.
- Ignorance: at the micro level most advice and information for industrialists, farmers and managers comes from financial consultants, chemical companies and the suppliers of industrial equipment. Ecological issues, and questions of environmental protection or public health, typically reflect short-term financial goals, rather than a vision of the whole environment from the point of view of the social interest.

> The major way in which knowledge is put to use is through commodities, but the commoditization of knowledge strongly influences the knowledge that is created . . . ignorance is not the passive absence of knowledge. Knowledge or belief about what? It takes on a particular form in Third World countries: developmentalism. Developmentalism holds that there is only one kind of progress, the conditions of the advanced countries of Europe and North America.
>
> (Haila and Levins 1992: 61–2)

Ignorance, then, is about not being aware of the alternatives. And in this context some of the most ignorant people are those who are the most highly trained: academics, consultants, scientists, experts.

Clearly, 'the environment' means different things to different people.

> The North Sea regarded as an instrumental resource is treated differently from the North Sea as a supply depot (for fish) or a habitat (for marine life) or as an amenity (for tourists).
>
> (Redclift 1987: 111)

Sustainable development

People live in different ways, and different theories *of* the environment reflect different ways of interacting *with* the environment. Indeed, the existence of various discourses on the environment is fundamental to this interaction, and if we are to understand and take action to rectify ecological degradation we have to understand the unexpected ecological effects of daily existence by being aware of the implications of alternative conceptions of the environment.

> [T]he 'environmental issue' . . . means such different things to different people . . . [t]hat a single word should be used in such a multitude of

different ways testifies to its fundamental incoherence as a unitary concept . . . The contemporary battleground over words like 'nature' and 'environment' is more than a matter of mere semantics, but a leading edge of political conflict.

(Harvey 1993: 1–2)

We are not sure whether a disaster is looming or not, and have no idea whether a policy of 'wait and see' may or may not make things worse. This uncertainty reflects a failure to appreciate why disasters are a potential: the natural effects of social activity. And where there is uncertainty or indeterminacy, then policy issues are based on beliefs. When politicians, consultants, scientists, academics or managers have to make decisions and necessarily have to theorize the environment, ambiguous or insufficient information requires that understanding is based on the *assumed* relation between people and the ecology (Adams 1995: 33; Holling 1979, 1986).

And assumptions are beliefs.

With regard to processes of sustainable development:

Each model fits the historical data equally well. The forecast is primarily determined by the *choice* of model, which in turn is determined by the forecaster's *assumptions* about the nature of the process he is attempting to predict.

(Adams 1995: 163)

Beliefs and assumptions reflect a broader world view, which in turn reflects an intuitive understanding by people of what are the 'viable' ways of organizing social life, intuitions that provide meaning to people's experience. And these interpretations of social existence become ideologies, which on the one hand help people to understand their experience and thereby make plans for their future (see Chapter 14), and on the other justify and legitimate people's lifestyles (see Chapter 18). These ideologies structure the way we understand life, and the way we discuss, recount and debate life processes: they set the parameters to theoretical discourse. And hence the answers 'appear to depend on who you ask' (Adams 1995: 50).

Both scientists and 'ordinary people' confront the world armed only with their myths of nature.

(Adams 1995:38)

Science and the myths of nature

'The environment' can *only* be conceived of theoretically. And *all* environmental theory is biased. Chapters 7–9 place the various debates and interpretations of the environment into a context that allows the implicit ideological implications of competing analyses to be identified, compared and contrasted. The intention is to allow you, the reader, to decide where you stand in the conflict, which is *what* the environment is, *how* it functions, and *why* it matters. There is no fence to sit on.

This follows the general theme of this book of: What *can* we know about existence (see the Prologue); and what *do* we know (see the Epilogue)?

> The scientific disagreement about the nature of the processes at work and how to model them, and the inability of scientists to settle their argument by appeal to the empirical data, provide a fertile environment for the development of biases. Biases, like mushrooms, flourish in the dark.
>
> (Adams 1995: 167)

Hopefully, Chapters 7–9 will throw some light on the environmental debate.

7 The environment as a source of pleasure
Egocentrism

The primacy of human wants: egocentrism

> [T]he value of the natural environment [is] in its instrumental role as a means of meeting human needs and gratifying desires.
>
> (Malnes 1995: 117)

Nature should be conserved for the sake of people. The natural environment is essentially inert and passive. Because humans are rational, sentient, independent beings, objective scientific knowledge can be utilized to dominate and bend nature to the fulfilment of people's subjective priorities. Individuals, because they are independent beings, their nature being biologically, genetically, inherited, should only be concerned about their own, unique, individual, needs. People are concerned to maximize personal utility according to their subjective preferences (see Chapter 3).

This is egocentrism.

Egocentrism defines relevant knowledge with regard to the environment as that knowledge that acts as an incentive for individuals to change their behaviour. And since humans are defined as independent, selfish beings, preoccupied with their own interests and pleasure, the incentives must be hedonistically expedient. Individuals *themselves* define relevant knowledge.

Hence, William Nordhaus of Yale University, musing on the effects of global warming on the USA, with implications for how far the USA (and other industrialized economies) should attempt to moderate the production of greenhouse gases as a by-product of the maintenance of their privileged lifestyles, can avariciously argue:

> Greenhouse warming would have little effect on America's national output. About 3% of American GNP originates in climate sensitive sectors such as farming and forestry. Another 10% comes from sectors only moderately sensitive – energy, water systems, property and construction. By far the largest 87% comes from sectors, including most services, that are negligibly affected by climate change ... In sum, the impacts of climate change on developed countries are likely to be small,

probably amounting to less than 1% of national income over the next half century.

(William Nordhaus quoted *The Economist* 7 July 1980)

Fundamentally, environmental protection is a question of conservation, reflecting individuals' preferences for consumption, which change as relative prices adjust to the forces of supply and demand. Environmental problems will essentially be experienced as shortages, which in free markets implies that prices will rise. And apart from individuals economizing on the use of relatively expensive sources of pleasure (or utility), there will also be an incentive for producers to find new sources of supply, and/or develop cheaper alternatives.

> *Any* system will produce destructive environmental impacts if the incentives within the system are not structured to avoid them. We have to look more deeply into any economic system to understand how its incentive systems work and how they may be changed so that we can have a reasonably progressive economy without disastrous environmental side effects.
>
> (Field 1994: 5, emphasis in original)

As we saw above (Chapter 3 – The subjective preference theory of value), that Barry Field can confidently assert that the activity in *any* economy reflects incentives to individuals influencing their consumption, is a distinct approach to economics in general, and to environmental economics in particular. This perspective is based on a set of restrictive assumptions about the nature of economic behaviour and beliefs about individuals' motivation, which are rarely justified as an analytical approach to human behaviour. It is assumed, *believed*, that individuals' potentials are biologically determined: people are endowed with a unique set of tastes and talents. And individuals need to be free to fulfil their genetically defined potentials – a freedom found in competitive markets (see Chapter 3). Without any justification, Field asserts that this is '*the* economic approach' (Field 1994: 4–5, emphasis added), not *an* economic approach.

David Pepper has characterized environmentalism that is concerned with the environment as a source utility (pleasure) for individuals as 'shallow ecology' and 'egocentric' (Pepper 1996: 41). Sometimes this approach is labelled *anthropocentric* – human centred (Fox 1990, Chapter 1) – but this does not acknowledge the importance of the subjective preferences of independent individuals in explaining choice and economic activity, which is crucial if the moral, ethical and ideological implications of this mode of analysis are to be identified and appreciated (see Chapters 15 and 18).

> Humans are acknowledged to be the sole reference point of value. They are what confers 'value', 'rights', obligations and moral duty, and they decide what is and is not to be valued. Human concerns are to be met by *using* nature.
>
> (Pepper 1996: 35, emphasis in original)

Within this perspective on the nature of human–environment relations, with regard to the future environmental viability of human life, there are the pessimists, forecasting impending catastrophe, and the optimists, who see no threat to the viability of life as we know it.

The pessimists

> If the present trends in world population, industrialization, pollution, food production and resource depletion continue unchanged, the limits to growth on this planet will be reached sometime within the next one hundred years. The most probable result will be a rather sudden and uncontrollable decline in both population and industrial capacity.
>
> (Meadows *et al.* 1974a: 23)

It is assumed (believed) that individuals are *only* concerned with their personal pleasure and consumption. Further, it is asserted that the earth has finite resources to meet human demands, and that population grows exponentially (2, 4, 8, 16, 32 . . .): consequently the carrying capacity of the planet will inevitably be exceeded. More economic activity by more and more people will mean that the growth in resource use, pollution and industrial output will be unsustainable, and will overshoot the earth's carrying capacity. In every resource category overuse and declines in quantity and quality were predicted: a conclusion endorsed six years later by the *Global 2000 Report*, commissioned by US President Jimmy Carter (GPO 1980).

These conclusions, popularized in the mid-1970s, were derived from The Project of the Predicament of Mankind, an initiative of the Club of Rome, which described itself as an informal, international association, dedicated 'to foster understanding of . . . the global system in which we all live . . . [and] to promote new policy initiatives and action' (Meadows *et al.* 1974a: 9). The project was a computer exercise in extrapolating past trends in population growth, agricultural production, natural resource depletion, industrial production and pollution into the future. Different assumptions, reflecting different scenarios, were fed into the programme – for instance increasing food production – and although this might delay the onset of global collapse, in each case, as one constraint is removed another is encountered. The collapse was predicted to begin in the year 2005.

Earlier, in 1968, Paul Erlich, in his book *The Population Bomb*, had asserted:

> [T]he causal chain of deterioration is easily followed to its source. Too many cars, too many factories, too much pesticide . . . too little water, too much carbon dioxide – all can be traced to *too many people*.
>
> (Erlich 1968: 66–7, emphasis in original)

For Erlich the population bomb was simply a biological problem, the process of human reproduction that has gone completely out of control. More recently

Paul and Anne Erlich, in *The Population Explosion* (Erlich and Erlich 1990), have revised the apocalyptic conclusions of The Project of the Predicament on Mankind, arguing that:

> Ending population growth and starting a slow decline is not a panacea; it would primarily provide humanity with the opportunity of solving its other problems
>
> (Erlich and Erlich 1990: 157)

But still the expectation is for 'the stork to pass the plow' (Erlich and Erlich 1990: 108): for the birth rate to surpass the production of food, resulting in mass starvation.

The *Limits to Growth* project of the Club of Rome was to quantify the apocalypse, detailing the effect of rising population and increasing industrial capacity consequent upon the growth dynamic of modern society. Ever scarcer resources would increase commodity prices, so that within a century world industry would be spending so much on ever more costly resources that it would be unable to renew depreciated capital. The industrial base would collapse, along with services and agriculture, causing a drastic drop in population as humanity returned to barbarism.

About the same time *A Blueprint for Survival*, endorsed by thirty-three leading scientists, with an introduction by Paul Erlich, and prepared by the editors of the *Ecologist*, called for population growth to be halved, and for economic growth to be slowed if 'the breakdown of society and the irreversible disruption of . . . life-support systems . . .' is to be avoided (*Ecologist* 1974: 16).

In general:

> Only by reducing and stabilising *all growth simultaneously* could ultimate collapse be avoided.
>
> (Pepper 1996: 67, emphasis in original)

This analytic approach is often called *neo-Malthusianism*, since in large part it follows the logic of the Reverend Thomas Malthus's *Essay on the Principle of Population*, published in 1798 (Malthus 1872). Thomas Malthus, a dogmatic man 'not to be intimidated by mere facts' (Chase 1980, quoted Pepper 1996: 179), was a passionate advocate of not relieving the plight of the poor: if they were helped to survive they would only breed, and the fundamental problem – that there were too many people to survive given the resources to produce food – would become worse, an analysis that prompted Marx to write that Malthus was not

> a man of science but . . . a bought advocate . . . a shameless sycophant of the ruling classes . . .
>
> (Marx, quoted Meek 1953: 123)

The trend towards population growth exceeding any possible increase in the production of food was a consequence of 'passion between the sexes', and the poor, who were less 'civilized' and consequently unable to exercise 'moral restraint' were essentially sexually incontinent. The poor were the 'problem'.

Neo-Malthusian thought is unable to conceive of qualitative changes in the way people socially exist – that is, changes in the way people socially interact to survive – resulting in improved technical relations in the production of the means of subsistence, and changed social relations: a more efficient technical division of labour. By focusing on independent individuals, it is a quantitative problem of the number of individual consumers. The myopia of the latter-day neo-Mathusians is a consequence of their methodology. Only quantitative changes in pollution, population, economic growth, industrial and agricultural production are considered (Meadows *et al.* 1974b). The qualitative basis of life, the *way* society organizes itself to collectively utilize the natural environment, is ignored.

> As Professor Kapp has pointed out, a computer set to work in 1872 would have predicted that the density of horse-drawn traffic would by now be so great as to render our cities impassable: choked as they would be with an impenetrable deposit of horse-dung.
>
> (Coates 1972: 105)

However, the apocalyptic predictions of the *Limits to Growth* thesis did not come to pass. And twenty years later a follow-up study, *Beyond the Limits* (Meadows *et al.* 1992), reinterpreted the *Limits to Growth* thesis. Although many pollutants had surpassed the levels that had been thought to be physically unsustainable, and consequently the potential for uncontrollable decline in food output, energy use and industrial output was still seen to be a potential,

> decline is not inevitable . . . Sustainability . . . is the ultimate challenge to the energy and creativity of the human race . . . we think the human race is up to the challenge.
>
> (Meadows *et al.* 1992: xvi–xvii)

The fundamental problem in achieving a sustainable lifestyle, such that the planet can absorb pollution and waste, and continue to provide clean air and water, food, raw materials and fossil fuels and not descend into chaos, is accurate data, for individuals to be 'aware' of what is happening. And although the earth has overshot the limits of sustainability, choices can still be made to bring

> . . . human society back from beyond its . . . limits . . . through human *choice*, human technology, and human organization, and these are *choices*

about . . . [t]echnological progress and market flexibility . . . [which can] bring the world to sustainability.

(Meadows *et al.* 1992: 8, 12, 9, emphasis added)

There must be reliable, adequate information that is timely, complete, and which is acted upon by individuals. And the parameters of the computer model upon which the *Limits to Growth* thesis had been based in the 1970s were updated to be more accurate. Information is the key to sustainability.

With different information structures, the system will inevitably behave differently . . . [and] only *individuals* perceiving the need for new information, rules and goals, communicating about them and trying them out, can make the changes to transform systems.

(Meadows *et al.* 1992: 223, emphasis added)

Such change will require: *visioning*, individuals realizing what they *really* want; *networking*, the dissemination of new information according to shared values; *truth-telling*, not distorting the facts for individual advantage; *learning*, action based on the new information; and *loving with a faith in humanity*, optimism based upon compassion. These are qualities that individuals, apparently, 'spontaneously' acquire. Human beings will develop 'new powers' within which 'each person will find his or her best role', which must be based on individual freedom since no leader 'no matter how authoritative he or she pretends to be, understands the situation' (Meadows *et al.* 1992: 224–33).

The assumption, the *belief*, is that individuals acting *independently* inexplicably hold the answer to our problems.

The optimists

The optimists do not have to rely on a mystical faith in the human character, which is spontaneously enlightened, with individuals exhibiting the cooperative qualities necessary for survival. The answer lies in the inherent selfishness of human nature and free markets: 'Self-interested behaviour . . . is *genetically* programmed into humans and is therefore inevitable' (Pearce and Turner 1990: 17, emphasis added).

The basic general argument is that the present form of economic expansion poses no threat to the viability of the environment but, on the contrary, contains the key to the solution to any environmentally related problems.

(Mellos 1988: 43)

In response to the pessimistic conclusions of the Club of Rome report, *Limits to Growth*, the right-wing think-tank, the Hudson Institute, commissioned a report, *The Resourceful Earth*, (Simon and Khan 1984), which concluded not only that population growth was not a liability but that rising population is an asset, but *only* as long as individuals enjoy the freedoms consequent upon a free-market, capitalist economy. And the secret lies within the inequality created, both within and between nations, which is a necessary consequence of market competition.

> The increasing disparity between average incomes in the richest and poorest nations ... we view this gap as the basic 'engine' of growth ...
>
> (Khan 1979: 60)

For Julian Simon the ultimate resource is the human mind (Simon 1981), which responds to new information in response to incentives – incentives that potentially augment individuals' pleasure (or utility). As we saw above (Chapter 3), where individuals are assumed to be independent beings, incentives that maximize utility presuppose free markets, and the environment, to maximize utility, should be conceived of as a commodity, traded according to consumers' subjective preferences just like any other commodity.

> The free market and political freedom are inextricably linked ... and they are based on respect for, and the right to private ownership. Ecological values can find their *natural* space in the market, like *any other* consumer demand.
>
> (Milton Friedman, quoted Ravaioli 1995: 32, emphasis added)

The environment as a source of pleasure (utility) will be *naturally* valued according to the enjoyment of consumers, but only in as far as there are well-defined property rights.

> Free market environmentalism emphasized an important role for government in the enforcement of property rights. With clearly specified titles ... market processes can encourage food resource [and natural resource] stewardship. It is when rights are unclear and not well enforced that over-exploitation occurs.
>
> (Anderson and Leal 1991: 3)

Where resources are a potential source of utility (an income stream from the sale of resources) to particular individuals, who are in competition with everyone else to maximize their pleasure (to achieve the highest selling price), then those resources will be managed very carefully to maximize potential income. As individuals compete to earn an income and consume, through markets, property rights will evolve, such that the environment, as so many sources of utility, will be traded as so many commodities.

This argument has been most vigorously advanced in the context of the Tragedy of the Commons debate.

> ... overgrazing of pastures and pollution of the oceans and the atmosphere result from the fact that common land, seas and air *are not owned*. If they were, then the resolution of damage levels and payments would be organized through the valuation and enforcement of the relevant property rights.
>
> (Helm and Pearce 1991: 8, emphasis added)

The debate was started by Richard Coase (1960). His concern was the differential between the private and social costs of production. Consumers pay a price for a commodity, reflecting the utility enjoyed in consumption (see Chapter 3), which compensates the producer for the costs of production. And an economy is organized to maximize individuals' utility when every individual is compensated for the disutility of production: labour is rewarded for its efforts, and displeasure arising from externalities, such as the effects of pollution, is treated as a cost of production. For example, smoke emissions from a factory are a disutility reducing individuals' pleasure that is not marketed and not reflected in the price paid by the consumer. And if market prices are to maximize independent individuals' utility, then these social costs have to be accounted for.

According to Coase the most efficient solution is a bargaining process between polluter and sufferer. If, legally, property rights lie with the polluter, then the sufferer could compensate the polluter *not* to pollute. However, if the sufferer legally enjoys property rights, then the polluter would compensate the consumer for tolerating the disutility of pollution.

Coase assumed markets to be perfectly competitive, which means that market prices accurately reflect the utility enjoyed (or disutility suffered) by individuals as consumers. And according to the logic of this analysis:

> *regardless of who holds the property rights, there is an automatic tendency to approach the social optimum.* This finding is known as the 'Coase theorem' ... *If it is correct, we have no need for government regulation ... the market will take care of itself.*
>
> (Pearce and Turner 1990: 72–3, emphasis in original)

Of course, there is no guarantee that the appropriate property rights will enforce an equitable reconciliation between consumers and producers, any more than some form of political regulation will achieve an 'optimal' level of pollution. Effective regulation presupposes omniscient, benevolent experts, able to accurately model ecological systems *and* predict individuals' preferences in consumption – and determine the appropriate incentives to change behaviour and achieve a solution.

... to assume experts will do what is 'right' rather than what is politically expedient is naive. There is simply too much evidence that politicians and bureaucrats act in their own self-interest ... '

(Anderson and Leal 1991: 170)

The 'commodification' of the environment into sources of utility to be traded might not be ever fully achieved, but for these theorists, given their beliefs in (assumptions about) human motivation, free market exchange offers the *only* possibility for ecological improvement *and* an expansion of individual liberty: the fundamental requirement of *all* social policy for believers in humans as innately selfish, utility maximizers (see Lal 1983, Chapter 1 and Chapter 7).

The ideology of free-market environmentalism

As we shall see in Chapter 15 this conclusion is crucial in the ideological justification and legitimation of economic inequality and the social status quo in market economies. The theory of perfect competition is not meant to describe any economy. Rather, if certain assumptions are made about economic motivation (that is, if certain beliefs about human nature are accepted), then perfect competition is the image of utopia, and economic policy is intended to make the real world more like the ideal, maximizing consumer choice. Hence criticisms of perfect competition in general, and the Coase theorem in particular, which dismiss these theories for not being 'realistic' (e.g. Pearce and Turner 1990: 73–8) completely miss the point of the analysis. On the theoretical confusion of the second best and the capital controversy see Chapters 3 and 4.

The importance of the theorem is ideological. The environment is reconciled to the economy, by using the institutions of the latter to achieve the goals of the former. Environmental protection can be achieved with the same, assumed, motivations that lead, theoretically, to economic growth and development.

That is a solution that unleashes the tremendous power of *self-interest.*

(Block 1990: 281, emphasis added)

The property rights argument, that private owners will act to obviate deterioration of the environment because they will reap the benefits of so doing (or suffer the consequences), was subsequently applied to analysing the viability of common property in development. Garret Hardin (1968) argued (an argument developed by Taylor 1976, and Laver 1981), that where there is common land upon which herders can graze their cattle, assuming that individuals are motivated to maximize utility, then each herder will graze as many cattle as possible. And when the carrying capacity of the land is reached, 'the inherent logic of the commons remorselessly generates tragedy'

(Hardin 1968: 1244). Because the benefits of grazing one more animal on the commons go to the individual herder, whereas the negative costs are socially felt by all the herders:

> the rational herdsman concludes that the only sensible course for him to pursue is to add another animal to his herd. And another, and another . . . But this is the conclusion reached by . . . every herdsman . . . Therein is the tragedy [over-grazing and despoilation of the environment] . . . Ruin is the destination toward which all men rush, each pursuing his own best interest . . .
>
> (Hardin 1968: 1244)

Hardin saw wider implications regarding the carrying capacity of the earth, with population growth exceeding the earth's capacity, 'and if this growth is not controlled the ultimate tragedy will be upon us – genocide' (Johnston 1996: 137) – a return to the *Limits to Growth* thesis.

And if we assume (believe) that individuals are *only* intent on maximizing personal utility, then this is the inescapable conclusion. But as we shall see in Part IV, this is only one possible conception of human nature, and beliefs in human nature have an ideological dimension, in justifying a particular social order with the claim that it accords with human nature. If individuals are believed (assumed) to be selfish utility maximizers, then a market economy is justified, and the rich and privileged *deserve* to be advantaged (see Chapters 3, 15 and 18).

It is a belief in individuals as independent, utility maximizers that is the dominant ideological belief in contemporary market societies, and which justifies economic and social inequality and the political status quo.

For Leach and Mearn, Garret Hardin confused common property with open access. Typically, social practices and local institutions that facilitate cooperation in the management of common resources evolve in the course of social life. Here the belief is that individuals are not independent beings, but depend upon each other for their survival, and people's intuitions, motivations and behaviour reflect the social parameters of existence (Leach and Mearn 1996a: 13; Bromley and Cernea 1989; Bromley 1992; Ostrom 1990; Shepherd 1989).

The Tragedy of the Commons logic has been applied to pollution in general: however, instead of taking resources *from* the environment it is the equally ruinous logic of excessive putting of pollutants *into* the environment. For the profit-maximizing firm wishing to dispose of pollutants, the cost of releasing waste into the environment is less than the cost of treatment, and for the individual consumer the cost of recycling is greater than the benefit. Social benefits and costs are not reflected in the decisions of individuals to maximize utility (and profit). This is just another Malthusian metaphor for population pressure on resources, which is clearly valid as long as enterprises are profit maximizing, responding to the preferences of consumers.

In considering the costs (to the sufferer/consumer) and the benefits (to the polluter/producer) of pollution, to be resolved by a bargain between consumers and producers, such that, under conditions of perfect competition, 'there is an automatic tendency to approach the social optimum' – the Coase theorem, see above – it is possible to conceive of an *optimal* level of pollution: that is, the level of pollution where the polluter's benefit (for instance, increased profit) from despoiling the environment just equals the sufferer's disutility from living in an inferior environment.

This optimum cannot be zero. Pollution is an incidental effect of an economic activity that has positive value. Commodities are produced that have a price, and are therefore valued by consumers. However, depending on the definition of 'property rights', compensation has to be paid either to the polluter to clean up their economic activity; or to the sufferer for the disutility endured.

> *This economic notion of an optimum has no necessary relationship to any biological notion.* The situation at the optimum may be wholly unsustainable . . .
>
> (Bowers 1997: 45, emphasis in original)

However, where the environment is addressed as a source of utility for consumers, the optimum can only be approached through market exchange. Positive benefits from pollution control have to be traded off against the negative impact on material wealth of either ceasing production, or the process becoming more costly owing to the additional costs incurred by pollution control. There is a trade-off between the social costs of pollution and any limitations on individual choice and therefore on utility enjoyed (and profits earned). And for each individual the trade-off will be different, reflecting individuals' unique tastes and talents and therefore subjective preferences: preferences that are revealed in the prices that consumers are prepared to pay for commodities.

It may be objected that by these criteria the rich would be free to pollute, as long as they pay for their 'pleasure'. But as we saw in Chapter 3, for free marketeers, according to their assumptions, in a perfectly competitive economy individuals receive an income in accordance with their social worth: the rich deserve to be rich. And where perfect competition does not obtain (and it never does), the *only* rational policy for governments is to improve the economy, making it 'more perfect', and not to redistribute income from those who have received more than they are worth to those who have been undervalued (see Chapter 3, and Lal 1983, Chapter 1).

The value to individuals of environmental amenities and resources can only ever be known by individuals themselves. Individuals' tastes for utility depend upon their unique, innate endowment of subjective preferences, which *can only* ever be revealed by their choices in free markets. Indeed, resources themselves are defined with respect to individuals' preferences.

Estimates of reserves at any moment of time never represent true resources in the sense of being all that can ever be found . . . the known reserves represent the reserves that have been *worth* finding, given the *price* and the prospects of demand and the costs of exploration.

(Beckerman 1974: 218, emphasis added)

As noted above, environmental problems will essentially appear as shortages, and as prices rise, consequent upon excess demand (see Chapter 3), consumption will fall back, and there will be increased incentives for producers to find new supplies, or to develop alternatives.

The neoclassical exegesis

The metaphysical belief, deduced 'from a few axioms . . . . [held] to be true a priori' (Shand 1990: 8), that individuals, motivated by hedonistic expediency, will spontaneously arrive at a bargain, so that in market exchange both parties (producers and consumers) will be better off, is associated with the Austrian School of economics.

Fundamentally the approach is subjectivist: the foundation of knowledge about the world is the private experience of individuals.

In fact most of the objects of social or human action are not 'objective facts' in the special narrow sense in which this term is used in the Sciences . . . So far as human actions are concerned things *are* what the acting people think they are.

(Hayek 1979: 44, emphasis in original)

Essentially, markets are only sources of information so that individuals can learn, given the preferences (tastes and talents) of other individuals, what are the potentials for their own pleasure.

Neoclassicism, a more 'orthodox', mainstream analysis of economic behaviour, though still oriented towards maximizing the utility of consumers, has pretensions to being a science that

studies the disposal of scarce means . . . Economics is the *science* which studies human behaviour as a relationship between ends and scarce means which have alternative uses.

(Robbins 1984: 16, emphasis added)

As we saw in Chapter 3, the concern is with the efficient allocation of resources to alternative productive uses, where 'efficiency' is defined by the maximization of individual consumers' utility, which is indicated by the relative profitability of enterprises and production processes. And positive statements can be derived that do not merely reflect what people *think* is the case, but what *is* the case.

It is possible to classify statements into POSITIVE statements and NORMATIVE statements. Positive statements concern what *is, was or will be*, and normative statements concern what *ought to be* . . . *disagreements over positive statements are appropriately settled by appeal to the facts.*
(Lipsey 1966: 4, emphasis in original)

Markets are still mechanisms through which individuals learn how to relate to other individuals in the maximization of utility, but markets also tend toward a definable equilibrium. And through a positive economic analysis, appropriate economic incentives can be purposefully defined and applied to influence individuals' choices to approach a social optimum: an equilibrium state where society's utility is maximized. For a mathematical derivation of a social, economic optimum, with regard to the use of environment see Pearce and Turner (1990), Chapter 18.

But where there is market failure – and markets are *always* imperfect – the social optimum does not obtain. Fundamentally this reflects inadequately specified property rights, necessitating institutional, economic intervention. Resources, for their 'proper' exploitation, must *belong* to someone.

Environmental pollution is a form of market failure, usually . . . when property rights are inadequately specified or are not controlled by those who can benefit personally by putting the resources to the most highly valued use.
(Pearce and Turner 1990: 17)

The theoretical logic and the ideological implications of this conception of economic experience are addressed in Chapters 3, 15 and 18.

With regard to the environment, neoclassical economists

seek to establish in the public mind a presumption that, once the institutional framework within which private enterprise functions, or ought to function, is extended to the management of the environment, the familiar mechanism of the market provides far more effective safeguards against environmental and ecological degradation than any extension of state controls.
(Mishan 1993: 223)

As we saw in Chapter 3, for utility to be maximized 'perfect' markets must obtain. However, a perfectly competitive economy cannot be empirically identified, and many environmental sources of utility are not marketed (for instance clean air). Therefore the 'correct' price signals, or *shadow* prices, to move towards equilibrium have to be theoretically defined.

One important way of achieving environmentally sensitive decision-making is to ensure that the environmental effects [of exchange relations] are better understood and that they are *valued in economic terms*.
(Pearce and Turner 1990: 224–5, emphasis added)

Cost–benefit analysis

Defining and applying economic incentives to vary individuals' choices to take into account possible environmental degradation, based upon valuing the costs and benefits of environmental sources of utility, is the most widely accepted approach to the economics of the environment: cost–benefit analysis (CBA). On examples of the use of CBA to inform political decisions on environmental questions and issues, see Barde and Pearce (1991).

> The task of environmental economics is ... to place valuations on environmental assets and consequences and, thereby, to develop appropriate policies.
>
> (Helm and Pearce 1990: 111)

The environment is conceived of as a basket of commodities.

> At heart, the neoclassical approach to environmental economics has one aim: to turn the environment into a commodity.
>
> (Redclift and Benton 1994: 69)

Market-based incentives, prices, are changed via taxes or subsidies, to alter individuals' demand for and supply of environmental commodities. Environmental protection through the manipulation of prices claims to be 'fair' since *all* people face the same prices, and value 'introduces flexibility into the compliance mechanism' (Pearce *et al.* 1989: 162). However, as we saw in Chapter 3, as an analytical approach it is plausible only if the beliefs of the researcher override analytical, logical inconsistencies.

The first paradox, or contradiction (depending on your point of view), is that a social valuation is derived from individuals' subjective preferences: individuals' preferences are held to be the 'proper basis for decisions about the environment' (Pearce and Turner 1990: 121). The valuing of such preferences is believed to encourage rational, economic decision making: where 'rational' is defined as 'profit-maximizing calculation'. The problem is that individuals' preferences are only ever revealed by the purchases they make in markets, which 'reveal' their subjective preferences.

> [V]alue only occurs because of the interaction between a subject [the individual consumer] and an object [the source of utility, the commodity] and ... is not an *intrinsic* quality of anything. A given object can ... have a number of assigned values because of the *differences in the perception* ... of human valuators ...
>
> (Pearce and Turner 1990: 212, emphasis added)

Value is *not* intrinsic, and is *only* revealed in the process of exchange.

Hence the 'social optimum' has no meaning outside individual consumer choice. The paradox/contradiction is how do we socially objectively quantify individuals' subjective preferences?

Within the techniques of CBA, expert knowledge is privileged over individuals' choices. Economists make decisions that will help *us* maximize *our* utility, without knowing our unique subjective preferences. And some of the techniques employed to find out what *we* think will be considered below. However, an underlying problem, which threatens the relevance of the whole CBA endeavour is: if individuals are unable to value, in monetary terms, say, global warming, saving the whale, or traffic congestion in our cities, what is the possible relevance of economists' surrogate valuations? And even if individuals can meaningfully conceive of the value of the environment in monetary terms, how can economists ever know what this valuation is?

> There is reason to suppose that such numbers [valuations] are meaningless abstractions even for economists who have professional interest in their being meaningful.
>
> (Adams 1995: 105)

See Adams (1995), Chapter 6, and below.

Accepting the relevance of CBA for the time being, the principal concern is to appraise public sector policy and investment decisions that are not subject to market forces. The anticipated social benefits (merely conceived of as the sum of individuals' preferences) from the policy/investment are compared with the social costs (both costs and benefits are estimated in value terms), and if the present value of benefits exceeds the costs, then the policy/investment is viable (on the issue of 'present value' see below). Where some of the costs/benefits are external to the society (e.g. pollution), normally, unless there are international treaty obligations, these are ignored.

There are a number of possible techniques to establish a social valuation based on supposed individual preferences, including expert opinion, cost of replacement, surrogate markets, travel cost, hypothetical markets, stated preference, and contingent valuation (e.g. Bowers 1997, Chapter 11, Pearce and Turner 1990, Chapters 9 and 10; Pearce *et al.* 1989, Chapter 3; Markandya and Richardson 1992, Part 2). Whatever the preferred method for valuing the environment, the aim is to quantify the *present value* of all future net benefits of the project, and to compare this with the capital investment, or to derive an *internal rate of return*, the rate at which the project provides benefits over and above costs, and to compare this with the rate of interest, and therefore the cost of the capital invested.

If an investment is expected to provide benefits over and above any costs (such as labour costs or the cost of raw materials) and negative side-effects (externalities that are disutilities) such as pollution, over the next five years of £5000 per annum, then we have to calculate what those future benefits are worth today. Benefits enjoyed in future years are discounted to an equivalent present value today. The discount rate, r, is equivalent to an interest rate: £1 invested at interest rate r will be worth £1$(1 + r)$ in year 1, £1$(1 + r)^2$ in year 2, £1$(1 + r)^3$ in year 3, etc. r is expressed as a decimal, so 10 per cent = 0.1. So £1 today, if invested, increases in value (what economists call *time preference*). So *today's* value is lower. Similarly, a return next year is equivalent to a lower value today. Future benefits are discounted. We shall deal with the choice of discount rate below, but the calculation of the present value (PV) of net benefits over the next five years will be as below:

$$PV = \frac{5000}{(1 + r)} \text{ [year 1]} + \frac{5000}{(1 + r)^2} \text{ [year 2]} + \frac{5000}{(1 + r)^3} \text{ [year 3]}$$

$$+ \frac{5000}{(1 + r)^4} \text{ [year 4]} + \frac{5000}{(1 + r)^5} \text{ [year 5]}$$

If the present value of the discounted future net benefits exceed the capital cost of the investment/project then the project should go ahead.

> [A] rational *social decision* is one in which the benefits to society (i.e. the sum of the people in society) exceed the costs.
>   (Pearce 1983: 3, emphasis added, parenthesis in original)

Alternatively an internal rate of return can be calculated. This is the discount rate such that the present value of the stream of net benefits just equals the capital cost of the project. This gives the actual, annual, financial rate of return of the project, for instance cleaning up beaches from sewage outfall.

So the present value of a future benefit is the economists' estimate of the sum of money that would have to be invested now, at the current rate of interest/discount rate, to produce a sum of money equal to the expected benefit at a future date. In terms of decisions made today, clean air in the future is less valuable than clean air today. And here is a moral issue: today's valuation of individual subjective preferences of future generations in order to maximize *our* utility. Today's winners value the loss of tomorrow's losers, conceptualized in terms that may have no relevance at all in a future society (even if we assume they are relevant to contemporary society).

It may be 'economically *efficient* for us to short change the future', but, depending on your ethical principles, *immoral* because it 'would amount to an unjust treatment of future generations' (Goodin 1992: 67, emphasis added). The correct valuation of long-term environmental protection relates to the efficient use of this generation's resources, and decisions concern the reassignment of resource rights to future generations. The moral basis of social, development, environmental and economic theory will be addressed in Part IV and Chapter 18.

For CBA to provide a basis for the allocation of resources to alternative uses, logically all the possible uses for resources should be valued, and those uses with the highest net present value or internal rate of return should be chosen.

[T]he logical implication is that *economic development* be measured so as to include changes in environmental assets and quality, and that investment decisions take environmental quality into account quite explicitly ... involv[ing] ... *all* the things that impact on individuals' well being ...

(Pearce *et al.* 1989: 37, 48, emphasis added)

Of course, such a comprehensive valuation of the possible use of environmental resources is never attempted, but as we saw in Chapter 3 the ideal policy formulation of individuals competing to maximize their personal utility is meant to be just that – an ideal. And it cannot be logically demonstrated that a partial improvement in some individuals' enjoyment of utility implies an absolute (social) improvement. This correlation is assumed: it is a belief and not a fact.

Cost–benefit analysis, then, claims to value the environment meaningfully, and to identify the 'best' use for resources, and in so doing to maximize the enjoyment (utility) of individuals, who are selfishly concerned with their own pleasure, and are essentially 'consumers'. However, as was discussed in Chapter 3, the analytical approach based on independent individuals maximizing utility through competitive exchange is concerned *only* with individual choice, and not with *social* choice: the paradox/contradiction of the social measurement of individuals' subjective preferences. Society is merely conceived of as the sum of the individuals that compose it, and social welfare is simply the sum of all independent individuals' utility satisfaction. Thus, social welfare would be improved if a few people became richer, more than the rest of society became poorer.

CBA has *absolutely nothing* to say about distribution. As long as the benefits, which might accrue to one group of people, exceed the costs, borne by a different group, then there is a net benefit to society, even though those who gain do not compensate those who lose. Now we can see the role of cost–benefit analysis as ideology, 'scientifically' justifying a particular allocation of social resources to the advantage of a particular individual or group of individuals. Positively ignoring distribution is not a failure of neoclassical economics, but illustrates its status as a science, as an ideology of knowledge – see Chapters 3 and 15.

While neoclassical theorists justify their analysis as positively reflecting what *is* – in a free enterprise competitive economy, economic transactions can be interpreted as merely reflecting individuals' subjective preferences for utility (see Chapter 3) – CBA analysts, working with shadow prices, are defining what *should be*.

What CBA produces, and what is *morally* correct, may coincide if, and only if, we adopt a further rule, namely that some *aggregated set of preferences of individuals* is the morally correct way of making decisions.

(Pearce 1983: 3, emphasis added)

For CBA to be morally legitimated, in the light of individuals' purchasing power, and hence the weighting of relative costs and benefits so that the analyses reflect the distribution of income, we have to believe that

> the existing distribution of income reflects the distribution of effort in the economy and . . . people deserve a 'proper' reward for their effort.
> (Pearce 1983: 7)

The rich deserve to be rich.

This is clearly an ideological issue and a political decision. But where there is objection to the distribution of resources affecting the outcome of political decisions to manage the environment – that is, individual's rewards *do not* reflect relative effort – then it is entirely possible to vary the weighting of costs and benefits to imply a fairer society (Pearce 1983, Chapter 5). Costs and benefits would be differentially valued according to *who* is advantaged *and* who suffers. Clearly such a procedure is replete with problems of analysts' obtaining 'appropriate' information, and the definition of what is or is not appropriate.

Social analysis in general, and economic analysis in particular, is evaluative: justifying a particular organization of society or allocation of resources. The logical impossibility of defining a social optimum in value terms, and of even deriving a meaningful internal rate of return, was addressed in Chapters 3 and 4. And as we shall see in Chapter 15, the ideological power of neoclassical economic analyses in general, and CBA in particular, is in the implicit justification of social inequality. Because market exchange is assumed (i.e. believed) to reflect individuals' subjective preferences and their talents and attitudes towards hard work, the resultant distribution of income is natural, and as long as there are free markets the poor have only themselves to blame: 'the distribution of income reflects the distribution of effort'.

The more that cost–benefit analyses are politically scrutinized according to *who* gains and *who* loses, detailed, context-specific information and data will be required, and there will be an unresolvable debate over what evidence is admissible: which 'is the main reason for not pursuing explicitly weighted procedures' (Pearce 1983: 70). The more market-based analyses are open to political scrutiny the less powerful they are.

Justifying a new road is a classic case of CBA as ideology. Those who enjoy the benefits of new roads, principally motorists, are rarely the same people as those who suffer the costs of noise, loss of amenity value of the countryside, houses and homes being demolished, etc. And the scientific economist, through CBA, may 'prove' that the net social benefit is positive, but those who suffer the costs

> . . . are not often content with the knowledge that other people will gain more than they will lose. And attempts to compensate the losers from

the gains of the winners routinely founder on disagreements about the valuation of the losses.

(Adams 1995: 171)

The valuation of a home, or the countryside, like obscenity, is in the eye of the beholder. It is a question of *individual* preference. While the market value of a home may be estimated by an estate agent, defining the compensation due for its demolition for the new road, for personal reasons the householder might feel that no sum of money would compensate him/her for loss of social networks and cherished surroundings. Many people resist the idea that such losses can be translated into cash at all. And if they are considered priceless, then the economist is obliged to enter this value into the CBA spreadsheet as infinity. And the present value of future costs and benefits is literally incalculable.

> [I]t only takes one infinity to blow up a whole cost–benefit analysis.
>
> (Adams 1995: 171)

So the question is redefined. Instead of trying to find out the compensation required for the loss of a home, the individual maintaining their home is now classed as a *benefit*, and people are asked what they would be prepared to pay for the status quo: the privilege of keeping their home! Such responses are within people's budgets, generating a meaningful number for the cost–benefit analyst (Mishan 1971). In the redefinition of the valuation of the property it is implicit that the individual has no right to the maintenance of the status quo: it is a benefit that has to be paid for, not a cost to be borne by the developer. And this political sleight of hand is 'scientific' (Adams 1995, Chapter 9).

> [A]sking people how much they would be prepared to pay to prevent a part of their birthright being taken away . . . is a form of blackmail – like the probing of an extortionist trying to find out how much a supermarket owner might pay not to have the goods on his shelves poisoned . . . rape preceded by cash compensation willingly accepted is indistinguishable from prostitution.
>
> (Adams 1995: 110)

The issue is, with regard to environmental degradation: are we to be compensated for the damage, or should we pay to prevent the damage? Is the economic regime to be permissive with regards to free enterprise development, or restrictive, with a premium on social lifestyles? The former is taken as the norm in cost–benefit analyses. In a guide published by the Department of the Environment of the British government in 1991, *Policy Appraisal and the Environment: A guide for government departments* (Department of the Environment 1991), preferences for change are distinguished from valuation of resources in themselves: CBA is concerned only with the latter. Once the question is framed within these parameters, the scope of any logical response is limited, and any other consideration is defined as incoherent and can legitimately ('scientifically') be ignored.

The agenda is also curtailed by the analytical methodology. Environmental commodities can be independently valued and conceptually removed from the ecosystem of which they are a part. The environment is a series of discrete, independent entities, and change is a number of independent 'events' (choices) in the exchange of commodities between consumers and producers. The environment cannot even be *conceived of* as an organic system.

> All social science is evaluative, with evaluation being grounded in the conceptions of humans and their place in nature on which this science is built.
>
> (Gare 1995: 155)

To understand incredibly complex processes we have to isolate those variables that we believe to be pivotal. Those features that we think we can act upon to change the outcome have to be highlighted. We have to conceptually abstract out of the real world those factors that are important to our actual world: where the *real* world is shared with other people, and the *actual* world is the realm of our experience. To explain the actual world of our experience we have to begin from conceptions of humans and their place in nature that can plausibly explain the process at issue. These conceptions are *beliefs* about human motivation, and the room for manoeuvre in controlling the parameters of social existence. What are the variables that could be manipulated to achieve preferred objectives, and what should these objectives be?

This is a theme to which we shall return in Part IV.

The commoditization of the environment

Reducing a multidimensional spectrum of values (religious, political, moral, aesthetic, ethical, emotional, sensory) to money, to economic calculation, inevitably biases the analysis. In the contemporary world 'money . . . is the basic . . . language of *social power*' (Harvey 1993: 4, emphasis added). For theorists working within the intellectual parameters of alternative perspectives,

> to undertake to measure the immeasurable is absurd and constitutes but an elaborate method of moving from preconceived notions to foregone conclusions . . .
>
> (Schumacher 1973: 42)

The analysis presupposes a society based upon commodity exchange. Environmental issues are represented as profitable enterprise: commodities that can be exchanged with respect to private property rights. It is impossible even to conceive of the interrelation between, say, rain forests, the upper atmosphere, plankton in the oceans, the whale population, etc.

Within these restricted intellectual parameters, the neoclassical theorist defines a social optimum: the point at which the cost and benefits of any activity are equated. Consequently, as we have seen, an optimal level of pollution can be defined: *physical* pollution is not the same as *economic* pollution. And there is no requirement for the sufferers of the disutility to actually be compensated by those who potentially benefit from this externality; it is enough that there is a net gain to society (to *other* individuals).

With the definition of an optimal level of pollution, pollution itself can be conceived of as a commodity to be traded. Permits for producers to pollute up to a certain level can be issued. For instance, the 1977 Clean Air Act in the USA allows producers to pollute the atmosphere up to a level considered by the political authorities to be an optimum. It then becomes economically rational for producers to clean up their production processes, so that atmospheric pollution falls below the threshold for which they are licensed, and then to sell the excess quota to other, less efficient producers. And in 1985 oil refineries were given quotas for lead in the run-up to phasing out the metal in petrol over a two-year period. In the early 1990s there was a trade in permits for power stations to pollute, particularly with regard to the emission of sulphur dioxide (SO_2), which causes acid rain. As power stations invest in flue gas desulphurization equipment or use expensive low-sulphur coal, thereby cutting SO_2 emissions, then these unused permits, or 'credits', can be sold at about $100 a tonne of SO_2, and operators are fined $2000 a tonne for excess emissions. It is claimed that the reduction in sulphur dioxide pollution in the USA from power stations between 1980 and 1996, by nearly a half, was a consequence of such financial incentives.

Commoditizing pollution through trading permits can be seen as more than a question of what should be the permissible level of emissions, and the enforcement of licences. Placing the issue of pollution within the intellectual parameters of commodity exchange legitimates competitive market processes. Questions cannot be asked *about* prevailing relations of exchange since this sets the terms of the debate. Competitive markets are effectively justified by not being questioned. Yet for many analysts it is the market rationality of exchange that is the fundamental cause of environmental concerns: see Parts I and IV.

As we shall see in Part IV and Chapter 18, the intellectual parameters of theoretical analyses reflect distinct social and economic interests – for instance at the eleven-day conference on the global environment held in Kyoto in December 1997, where diplomats from 159 countries, lobbied by a whole spectrum of special interest groups, discussed among other issues global warming. It was not until within 24 hours of the close of the conference that Raul Estrada, chairman of the conference, produced targets for the emission of greenhouse gases, thought to cause global warming. Straight away there were disputes between the industrialized countries over the suggested targets, although a deal was eventually struck that called on the USA to reduce emissions by 7 per cent by the year 2012. But the USA refused to agree to *any* target without trading in permits to discharge carbon dioxide into the atmosphere. The parameters of environmental control were set by market exchange, terms of social interaction that have a distinct ideological bias (see Chapter 18).

The insistence of the US negotiators on permit trading becomes clear in the light of the Russian economic crisis. With the collapse of the Russian economy and the consequent rise in unemployment and reduced industrial output, emissions of CO_2 have plummeted, and 3.7 billion tons of gas have been 'saved': permits that the USA offered to 'buy' for hard currency, allowing America to stay within the terms of the agreement and yet increase the emission of greenhouse gases! Enraged by the USA's being able to 'cheat' on agreed obligations, the G77 group of developing countries refused to sign any deal, and eventually a cap was placed on emissions trading of no more than 50 per cent of any binding targets.

The Economist (13 December 1997) considered the acceptance of emissions trading as 'a great leap forward in environmental thinking'. This enthusiasm will be explained in Chapter 15. In November 1998 representatives of the 167 countries that signed up to the Climate Change Convention met to discuss the details of the control of emissions. It was largely an agreement to do nothing. The important decisions on greenhouse gases were postponed for two years. Two weeks of horse-trading over permits left 142 unresolved items on the agenda. And it emerged that the World Bank planned to take control of the estimated $90 billion (by 2020) international carbon trading market, charging a 5 per cent commission on all deals, generating an income of some $4.5 billion. However, it was pointed out that there was a conflict of interest given that since the 1992 Earth Summit in Rio de Janeiro the Bank had spent twenty-five times more on fossil fuel power generating projects than on renewable sources of energy.

8 The environment as a productive resource
Ecocentrism

The dependence of humans on the environment: ecocentrism

For egocentric theorists, people are independent individuals who are naturally selfish, and the purpose of life is to maximize individual utility. The environment is conceived of as a source of pleasure: nature exists for the convenience of humankind. However, for many analysts it is precisely this attitude that has created the environmental problems of pollution, global warming, acid rain, toxins in processed food, and so on. And underlying such a hedonistic understanding of the man–nature relationship is the myopic view of humans as naturally selfish, independent beings – a theme we shall return to in Part IV. In consequence, it is the dominant belief system which has to be questioned. In this context, back in 1966, Lynn White described Christianity as the most anthropocentric

> religion the world has seen ... [bearing] a huge burden of guilt [for the] dogma of man's transcendence of, and rightful mastery over nature ...
>
> (White 1966 :25, 27)

Environmental problems are not to be solved by ever more technically sophisticated ways of using nature simply for our benefit: there has to be a fundamental rethinking of our attitudes and values about the place of humans in the cosmos.

> [T]he 'control of nature' is a phrase conceived in arrogance, born of the Neanderthal age of biology and philosophy, when it was supposed that nature exists for the convenience of man.
>
> (Carson 1962, quoted Fox 1990: 5)

Ecocentric thinkers do not subscribe to this 'Neanderthal arrogance', and are opposed to the trite simplicity of environmental economics and the myopic thinking of egocentric theorists, they are concerned with *ecology*, the

science of the relations of living organisms to the external world, their
habitat, customs, energies, parasites, etc.

(Worster 1985: 192)

They are concerned to understand the ecological parameters of human
existence. People are not independent beings, free to choose as consumers,
but they depend on the environment for existence. Relevant knowledge
is now defined *external* to the individual, rather than internally reflecting
individuals' preferences. And this dependence, the way we technically relate
to the environment to produce the wherewithal for existence, is managed
through social institutions. Ecocentrism implies a different belief in human
nature. The biological determinism of the egocentrics, which holds that
individuals' behaviour reflects their genetic inheritance, and the best that
policy makers can do is to harness these natural drives so we can each benefit
from each other's selfishness, is supplanted by a form of technical determinism. The technical exigencies of human survival institutionally shape the
social structure, and individuals are socialized to adapt to the prerequisites
of survival.

Individuals' behaviour has to be *eco*centric rather than *ego*centric.

This is a view that implies a degree of fatalism: 'nature's laws determining
events' (Pepper 1996: 246). For Zukar (1980) such parameters to human
existence make the concept of free will extremely limited: the best we can
do is define the ecological limits to existence, and learn to live within them.

Markets are no longer conceived of .as mechanisms through which
individuals compete to maximize utility in consumption; markets are now
institutions through which individuals cooperate to produce. And free markets
are ever less applicable as the technology of production becomes ever more
complex – a theme addressed in Chapter 4. This chapter is concerned to
identify and explain the various interpretations of ecological thought, where
individuals are believed to be dependent, socialized, beings (on the spectrum
of thought that lies within the perspective of ecocentrism, see Pepper 1996,
Chapter 1).

The technical production of social existence shapes society's and individuals' social experience. This experience is reflected in people's attitudes,
intuitions and motivations. Consequently, an environmentally benign society
is contingent upon

- a changed technical basis to production. For E. F. Schumacher, technology has to be 'appropriate' to the culture of society: 'If the purpose
  of development is to bring help to those who need it most, each "region"
  or "district" . . . needs its *own* development' (Schumacher 1973: 147,
  emphasis added). However, the World Commission on Environment
  and Development recommends that societies should adapt to the needs
  of advanced technology to promote economic growth and poverty alleviation: 'economic growth is seen as the only way to tackle poverty . . .

It must however be a new form of growth, sustainable, environmentally aware' (World Commission on Environment and Development 1987: 59);

- and/or the social management of production such that the excesses of individuals' preferences for consumption do not compromise the integrity of the ecological relation between humans, societies, and the environment;
- and/or a changed awareness in individuals, such that the technical, and hence social consequences of existence structure their behaviour – ecological awareness becomes an intuition.

The ideological implications of such an understanding of man–nature relations will be addressed in Chapters 16 and 18, but for the purposes of the argument so far a managing elite is justified. Expert objective knowledge is valued over the subjective preferences of independent individuals. The natural environment is a technical resource to be managed in the general interest. Because production implies technical specialization, people do not consume what they produce; they depend on each other and on the technical division of labour for survival. Individuals have to cooperate through social institutions; people are integral components in a survival *system*, and their activity must take cognizance of their dependence.

> Increasing awareness that our global ecological life support system is endangered is forcing us to realize that decisions made on the basis of local, narrow, short-term criteria can produce disastrous results globally and in the long run.
>
> (Costanza *et al.* 1991: 2)

Political direction and control through the state management of economies has been an inevitable consequence of the evolution of technology and higher living standards. And the new industrial revolution, based upon computing and telecommunications – the information revolution – which has eroded national boundaries, globalizing economies, is often thought to have rolled back the state as an economic force. But in spite of the neo-liberal emphasis on free trade unimpeded by state intervention (see Chapter 3), and the conservative ideological pressure to expand individual liberty at the expense of social responsibility and collective interest (see Chapter 15), average government expenditure as a percentage of gross domestic product across the industrialized world has increased from 27.9 per cent in 1960, to 42.6 per cent in 1980, 44.8 per cent in 1990, increasing to 45.99 per cent in 1996 (IMF figures, cited in *The Economist* 20 September 1997).

> [A]s societies became more complex and science and technology advanced humans encountered the need for greater controls on the environment.
>
> (Wells 1996: 8)

What form these controls might take – appropriate technology, efficient management, the education of people and the popular adoption of ecocentric beliefs – is the subject of debate. And it is the project of this chapter to try to understand this spectrum of thought and argument.

Policy pragmatists

This school of thought within the ecocentric perspective, while accepting that prices reflecting individuals' enjoyment of utility would be the best of all possible worlds, does not share the belief that individuals' behaviour will spontaneously so evolve, nor that the environment can be meaningfully valued to be included in the equation. However, economic instruments remain a tool, which the political authorities may utilize to regulate the environmental effects of human activity.

Problems arise first because of market failure, the economy being distorted by prices' not reflecting individuals' enjoyment of utility. And for the policy pragmatists, addressing environmental problems piecemeal – adjusting particular prices to affect particular markets rather than addressing economic imperfections in general – is a viable option (Munasinghe *et al.* 1993). As we saw in Chapter 3, such a belief has no grounding in economic theory, and is just that, a *belief*. For the pragmatists the problem lies in not being clear as to the complicated process by which macro (society-wide) level policy changes ultimately affect incentives for the efficient allocation of resources at the micro (individual) level. However, this is a difficulty that potentially can be overcome by more detailed empirical research on the implications of relations of exchange.

> Because the full consequences of a policy are not traced, both quantitative and qualitative results of the partial equilibrium model may be wrong.
>
> (Cruz *et al.* 1997: 2)

For other policy pragmatists the fundamental difficulty is in internalizing market externalities: that is, the problem of setting appropriate shadow prices to represent the non-marketed consequences of market exchange – essentially the problem of reflecting *social* costs as an exchange between *individuals*.

> [T]he existence of environmental *social* costs is a very strong argument for installing *social institutions* responsible for environmental quality and equipping such institutions with instruments that will enable them to reach *socially* desirable environmental objectives.
>
> (OECD 1989: 12, emphasis added)

Economic incentives remain secondary to other policy tools that enable a much more direct regulatory approach towards environmental degradation.

However, economic incentives, which stimulate particular types of economic behaviour, charges and subsidies and the definition of property rights, may have a role to play, but the difficulty is in defining 'the economic'.

> [T]he notion of an 'economic' instrument has come to mean different things in different contexts and as perceived from *different views of what economics is about.*
>
> (OECD 1989: 14, emphasis added)

We highlighted the problem of defining 'economic' behaviour and what 'economics is about' in Part I, and this need not be repeated here. But the point remains that for this school of thought the apparent flexibility of economic regulation is outweighed by administrative complexity.

> Economists are aware of all that is attractive about pricing mechanisms, and what is attractive comes out of economic theory, whereas administrators and legislators are aware of all the practical reasons why pricing would rarely work well enough for the theoretical virtues to outweigh the practical objections.
>
> (Schelling 1983, quoted OECD 1989: 127)

Uncertainty and the Gaia thesis

It is precisely the inability of theorists to accurately model relations between the environment and society, and the failure of economists to value these relations and to foresee the global (macro) effect of partial (micro) changes in the way people relate to the environment, that has led James Lovelock to popularize the notion of 'Gaia'.

> . . . Gaia, the Earth goddess, oldest and greatest of the pre-classical of the Greek pantheon of gods.
>
> (Lovelock 1991: 24)

The Gaia hypothesis was originally advanced by Lovelock and Margulis in 1974.

> At the time . . . most earth scientists thought of life . . . as an opportunist adapting to the global environment . . . [however] most scientists today would probably agree that it was the emergence of life on Earth . . . which caused our planet's evolution to diverge from that of Mars, Venus and the moon.
>
> (Watson 1988: 66)

Lovelock likens environmentalism to planetary medicine, which is based upon 'geophysiology, the systems science of the earth' (Lovelock 1991: 10). The earth is conceived of as a living organism to the extent that the chemistry and temperature of the planet are self-regulating. For this reason the environment has to be understood holistically. Whatever we do in one part of the global ecosystem will ultimately affect all the other parts, eventually reverberating on ourselves. And we are an integral part of this mutual dependence. The atmosphere, the oceans and climate are regulated to sustain life *because* of the activity of living organisms.

The Gaia hypothesis is *not* saying that the earth *is* alive, only that our knowledge of the global ecosystem is so primitive that the consequences of our actions can only be guessed at. The prescription may well be worse that the problem: 'prescribing remedies is quite beyond our skill and dangerous to attempt in ignorance' (Lovelock 1991: 174). We have to be aware of the limits of science, in particular: scientists are divided into various disciplines, such as meteorology, biology, marine science, and atmospheric chemistry, and their professional training militates against their conceiving of the whole. Scientists write 'in the secluded dialects of reductionist expertises' (Lovelock 1991: 27). We have to forget our particular social and intellectual concerns; we have to escape the myopia of human existence and concentrate on the eco*system*.

> In Gaia we are just another species, neither the owners nor the stewards of this planet. Our future depends much more upon a right relationship with Gaia than with the never ending drama of human interests.
>
> (Lovelock 1989: 14)

Rather than trying to 'manage' the earth we must learn to live *with it*. We have to regulate our own lives rather than attempt to regulate the ecosystem.

Social regulation and ecology

> Economics cannot on its own provide a full answer . . . [there are] broad social and moral issues that can only be resolved, if at all, by extended and wide-ranging debate.
>
> (Bowers 1997: 4)

Social existence in general, and economic activity in particular, has to be within social limits, but what these limits might be is an issue of heated debate: heated, because different understandings of the social basis of human existence have distinct ideological and political implications. All ecocentrics agree that the emphasis of egocentric theorists that free markets are the answer is at best simplistic, and most agree that such a response will make the ecology of the earth even less viable. But here the similarity ends.

Those theorists who come to study the environment from economics – the egocentrics – we can call *environmental economists*. The school of thought of environmental economics is very different from *ecological economics*.

ecological economists, *eco*centrics, begin the analysis from the ecosystem, not from economic behaviour. They take 'a wider and longer view in terms of space, time and the parts of the system to be studied' (Costanza *et al.* 1991: 3). With the combination of economics and ecology the evaluation of the environmental effects of economic activity is not now a *quantitative* cost–benefit analysis (CBA) exercise, involving the valuing of environmental commodities, with market exchanges being simulated through the derivation of shadow prices. Rather, an environmental impact assessment (EIA) is carried out: a *qualitative* technique developed in the USA as a consequence of the National Environmental Policy of 1969. EIA

> is essentially a technique for drawing together, in a systematic way, expert *qualitative* assessment of a project's environmental effects, and presenting the results in a way which enables ... them to be properly evaluated by the relevant decision making body ...
>
> (Department of the Environment 1988: para 7, emphasis added)

EIA does not have the scientific pretensions of CBA, and does not try to derive an optimum solution.

> ... as 'science' ... [EIA] has to do with the methodologies and techniques for identifying, predicting, and evaluating the environmental impacts associated with particular development actions ... as an 'art' ... [EIA] has to do with those mechanisms for ensuring an environmental analysis of such actions and influencing the decision-making process.
>
> (Kennedy 1988: 257)

There is not space here to go into the detail of the techniques of EIA, but see Wood (1995), Smith (1993), Turnbull (1992), Bartlett (1989), and Environmental Protection Agency (1993).

But if there is not the faith of the environmental economists that market processes will automatically vary individuals' behaviour to alleviate, or even regress, environmental degradation, how do ecological economists anticipate that change will come about?

While new technology might mollify the negative aspects of social existence on ecology, and 'science and its application, in the form of technical progress, allow for tremendous economic growth ... new institutions [will] have to be invented and innovated' (Faber *et al.* 1996: 66, 65) if people's attitudes, intuitions and behaviour are to become ecologically benign. We have to learn to temper the economic dynamic of social life, not only with scientific ecological knowledge, but also with an ethical dimension founded on ecological reasoning. Mankind has to realize that 'we are part of a unity with nature' (Faber *et al.* 1996: 255).

The integrative power of such an ethical awareness on economic and ecological existence can be created through a learning process ... How we learn and how we teach about things like the world environment which are not obvious in ordinary daily experience, is an important problem that environmentalists must give a great deal of attention to.

(Boulding 1991: 30)

Such an educational curriculum to generate environmental ethics should include:

- ecological principles for sustainable development;
- psychological factors for the creation of sustainable societies;
- a critique of orthodox economic theory;
- training to build appropriate institutions for local, regional, and global sustainability (Clark 1991).

For Herman Daly and John Cobb appropriate education is not enough: '[W]e believe that our collective response ... is above all a matter of religious conviction and vision' (Daly and Cobb 1990: 355). The analytical emphasis is the same: that people are dependent on two systems – the ecological and the economic – and the management of social life cannot be achieved merely by economic techniques. The vision has to be wider. People have to realize that economic and ecological problems are interconnected, 'indeed have a common source' (Daly and Cobb 1990: 356). People can be attracted to new ideas and ways of organizing their lives as they realize the disastrous consequences of living within the status quo.

The major obstacle is habits of mind. We have to change: social organization; educational and university curricula; trade policy; establishing economic and social life on the 'right scale'; economic progress must not be conceived of as economic growth ... *all* are important, and all are interlinked, and hence change will not occur without attitudinal change.

> ... the changes that are now needed in society are at a level that stirs religious passions. The debate will be a religious one whether that is explicit or not.
>
> (Daly and Cobb 1990: 375)

It is religious because it is a moral issue. Humans are not merely one species among many. We are *special*. How do we know that? Because the Bible says so. We are told (presumably by God) that we have dominion over all other creatures – the sentiment that Lynn White found so objectionable above, and Carson thought was born of Neanderthal arrogance – but for Herman Daly and John Cobb this ascribed status has been misunderstood. We are not entreated to exploit the environment for our pleasure, but to manage it responsibly as a resource. Rather than being classified as *eco*centric theorists, they are really *theo*centric thinkers.

Apart from God, we think, there would be no meaning, no life, no righteousness, no truth, no value . . . We believe that human life is lived most richly and most rightly when it is lived from God and for God.

(Daly and Cobb 1990: 394)

The theorist has to be the evangelist.

The *deep ecologists* (a term first coined by Arne Naess in 1973) similarly are waiting for a religious awakening.

Deep ecology goes beyond a limited piecemeal approach to environmental problems and attempts to articulate a comprehensive religious and philosophical worldview.

(Devall and Sessions 1985: 65)

Ecosophy as a philosophy of life however does not accept that humans have a special, guardian status in the cosmic order (Naess 1989; Devall 1988).

The central idea [of deep ecology] concerns the rights of every form of life to function normally in the ecosystem . . . rivers [have] a right to be rivers, mountains to be mountains, wolves to be wolves . . . humans to be humans.

(Nash 1989: 146–7)

Maintaining the richness of diverse forms of life becomes an *end in itself.* The awareness that people are not separate and isolated from the rest of nature will be a consequence of 'an active deep questioning and meditative process and way of life' (Devall and Sessions 1985: 66) – an awareness that becomes an intuition, which goes hand in hand with cultural change. We have to move away from the computer consciousness of contemporary technical/industrial culture. We have to minimize our needs to limit disturbance to the ecosystem. And we have to emphasize the quality of life rather than the quantitative rise in living standards. People must be aware of their vital needs and how these are related to the vital needs of all other beings. There is a need to reduce population to a sustainable minimum.

I should think we must have no more than 100 million people if we are to have the variety of cultures we had one hundred years ago.

(Arne Naess, quoted Devall 1988: 76)

The relationship to nature is *not* the 'dominion' of Herman Daly and John Cobb.

. . . [the] basic intuition is that all organizms and entities in the ecosphere . . . are equal in intrinsic worth.

(Devall and Sessions 1985: 67)

A school of thought that can broadly be called *social ecology* combines the insights of the deep ecologists with the Gaia theorists. Living organisms 'play an *active* role in their own survival and change' in a process of 'participatory evolution' (Bookchin 1990: 107, 108, emphasis added) through which they shape their own environment. And the realization that 'we are part of a "community of life"' (Bookchin 1990: 103) has to become an 'intuition'. The ecosystem is not there just for our convenience. If such an awareness does not develop then we are in danger of undermining the natural world upon which we depend for our survival, because we will try to find 'substitutes' so that we can plunder the environment to 'maximize utility', and 'it is precisely this view that has played a major role in the present ecological crisis' (Bookchin 1990: 103).

People have to do more than describe what *is*, and have insights into what *could be*. Intuitively, there has to be an awareness of the potentials of social change, which can only be identified through *dialectical naturalism*.

> Dialectical contradictions exists within the structure of a thing or a phenomenon by virtue of a formal arrangement that is incomplete, inadequate, implicit and unfulfilled in relationship to what it 'should be'.
>
> (Bookchin 1990: 32)

These contradictions are beyond mere reason; their discovery is also a product of feeling, sentiment and moral outlook. 'Our very *passion* for the truth . . . [must not] overlook *other forms of reason* that are organic and yet retain their critical qualities, ethical and yet retain contact with reality' (Bookchin 1990: 11, emphasis in original). As we shall see below the concept of the *dialectic* as used here is very different from the concept as used by Marxists, for whom it is not an approach to discover the truth, but a methodology for understanding change – change which is often unpredictable – and hence it is not a question of discovering how things *should* be.

But importantly the social ecologists go beyond religious conviction as the force to change society.

> The task is to create a political sphere that other movements have not seriously tried to do . . . calling for the development of a network of local assemblies which alone can constitute an authentic democracy.
>
> (Bookchin 1997: 17)

This will reflect a growing culture of public involvement in social affairs and change, an *ideological* change. The way we relate to and view the natural world derives from the way we view the social world, which is a product of social ideology (Bookchin 1990: 112). And social ecology

> takes on the responsibility of evoking, elaborating, and giving an ethical content to the natural core of society and humanity . . . the social

conceived of as a fulfilment of the latent dimension of freedom in nature and the ecological conceived of as the organizing principle of social development.

(Bookchin 1990: 117, 118)

Political ecology has a similar, if less specific emphasis, 'combining the concerns of ecology and a broadly defined political economy' (Blaikie and Brookfield 1987: 17). This combination, however, is extremely ill defined.

The task of the development of political ecological understanding of people and environment may be one of:
- identifying productive contradictions between and within its parent disciplines
- ordering these contradictions into a set of mutually interacting discourses
- providing a negotiating space for the contradictory views and approaches that politicized environmental studies requires.

(Blaikie 1995: 8)

The notion of a contradiction is understood as a *logical* contradiction, not as a *political* contradiction (see Chapter 17). Problems can be solved rationally through interacting 'discourses'. We are thought to be in, theoretically, confusing times. Understanding proceeds through scientific revolutions: we develop intellectual notions, or *paradigms*, as to how the world works so that we can manage our experience to achieve our ends (see Chapter 16). However, it is believed that periodically our paradigms are unable to rationalize our behaviour, and logical contradictions emerge that the extant paradigm cannot explain, and which hitherto we have not experienced. There is a period of revolutionary science. Intellectuals engage in feverish activity to understand life 'providing a negotiating space for contradictory views', views reflecting different aspects of this newly confusing experience. Eventually a new paradigm will emerge that will make sense of these conflicting (contradictory) experiences, and then we can settle back into a rationally, managed existence.

Nowhere do the political ecologists address their implicit theory of knowledge, which defines their ideological bias and is only *one* possible conception of science (which will be addressed in Part IV), and consequently the 'political' cannot be defined. And this failure to acknowledge their particular ideological bias effectively takes the politics out of political ecology. While 'conflict and contestation over the environment are central concerns of political ecology' (Blaikie 1995: 15), this conflict merely reflects a lack of understanding of *how* society and the environment interact: it is not a question of *why* the necessary interaction takes on a particular social form. On the methodology of 'how?' and 'why?' questions see Chapters 16 and 17. For the political ecologists conflicts can be resolved rationally and through the activity of intellectuals, usually in universities, 'discursively' analysing, debating, and theorizing how we live our lives. Expert knowledge is fundamental to environmental social change (Keil *et al.* 1998).

In this intellectual endeavour to understand social life, and therefore take action to preserve the environment, Western scientific thought to be apposite has to be counterpoised to local knowledge systems, so that individual experience can be addressed, whether that may be the experience of 'a mother, a wife, a trader, a fuel gatherer, or a grower of household staples' (Blaikie 1995: 17). In a very 'neo-populist' sense (see Chapter 12), the real world is the world as directly experienced by people: it does not have to be theoretically interpreted to reveal the social dynamic of individual behaviour (see Part IV). Reality is not problematic; we just need more observations of more people's social lives, and of the environment, and then need to discover the paradigmatic formula for reconciling these distinct areas of inquiry. The environment is composed of

> natural materials as transformed by particular people from particular positions in particular societies according to their own belief systems informed in turn by their experiences and needs . . .
>
> (Blaikie 1995: 4)

We need a methodology that 'emphasizes the constantly shifting *dialectic* between social and land based resources' (Blaikie and Brookfield 1987: 17). Again, the 'dialectic' *describes* the interdependence between the environment and people. This is very different from the way in which this concept is employed in Marxist analyses to *explain* the social context of people's existence (see Chapters 5 and 17). The use of 'radical' words (if not the concept) gives the illusion of a political radicalism that is wholly unwarranted.

> The theoretical heart of this body of scholarship [political ecology] . . . [is] the linking of political economy (typically of a Marxist or neo-Marxist variety) with ecology . . .
>
> (Peet and Watts 1996a: x)

However, the lack of an explicitly political agenda in political ecology has spawned *liberation ecology*, a school of thought intending 'a practical political engagement with new movements, organizations and institutions of civil society, challenging conventional notions of development, politics, democracy and sustainability' (Peet and Watts 1996b: 3). It is not just that poverty causes environmental degradation

> by the rational response of the poor households to changes in the physical, economic and social circumstances in which they define their survival strategies.
>
> (de Janvry and Garcia 1988: 3)

Degradation also appears to cause poverty, but poverty is part and parcel of deeper relations of social inequality:

> those who place an undue emphasis on poverty [political ecology] . . . must recognise that impoverishment is no more a cause of environmental deterioration than its obverse, namely affluence/capital.
>
> (Peet and Watts 1996b: 7)

It is a consequence of social relations between the rich and the poor, not the relation between the poor and the environment. Within political ecology there is an unspecified, deterministic relationship between the natural environment and people that produces 'an extremely diluted, diffuse . . . 'conjunctural' explanation . . . [and] actually presents an *ad hoc* and frequently voluntaristic view of degradation' (Peet and Watts 1996b: 8).

It will be argued in Chapter 18 that the definition of politics implicit within political ecology is valid only where politics is conceived of as achieving compromise in the common interest: an interest that is being redefined in the search for a new paradigm in these scientifically revolutionary times.

> While our sensitivity to the influence of social and political forces has certainly grown, our understanding of their actual impact on the production of scientific knowledge has not.
>
> (Keller 1985: 5)

9 The environment and social evolution
Sociocentrism

The co-evolution of society and the environment: sociocentrism

We saw in Chapter 7 that egocentric theorists understand the environment as a source of utility – as a means to satisfy individuals' preferences for pleasure. Ecocentric theorists however recognize that the natural environment functions and has a rhythm according to natural laws and processes that are beyond human control and which, if we are to survive as a species, we have to respect.

For egocentric theorists human existence is essentially independent of the natural world, and as long as society is organized through competitive, free markets, human behaviour will naturally adjust to the ecological limits to human existence. But for ecocentric theorists individuals are dependent (not independent) beings. We are *not* free to choose as consumers, but have to learn to cooperate within a technical division of labour, as producers. And the natural environment as an input into the production process has to be managed, or at least the natural, technical laws of sustainability have to be respected, if production processes are to remain sustainable.

However, for *socio*centric theorists, individuals obviously *do* choose between alternatives to consume according to their preferences, *and* they have to cooperate with others to produce and to survive within environmental parameters. It is not an 'either/or'; people's independence is conditioned by their dependence. People are *inter*dependent with each other, *and* with the natural conditions for their survival. Individuals in society, and the environment, define and determine each other in a dialectical, co-evolutionary process of change. The relationship is not dialectical in the simplistic sense in which the term is used by political ecologists, *describing* interdependence; but the relationship is dialectical in the *explanation* of change and evolution.

> For dialectics the universe is unitary but always in change; the phenomena we can see at any instant are parts of processes, processes with histories and futures whose paths are not uniquely determined by their constituent units. Wholes are composed of units whose properties may be described, but the interaction of these units in the construction of

the wholes generates complexities that result in products *qualitatively different* from their component parts ... In a world in which such complex developmental interactions are always occurring *history* becomes of paramount importance.

(Rose *et al.* 1984: 11, emphasis added)

We shall return to dialectics as an analytical methodology to understand *inde*terminacy, processes 'resulting in products *qualitatively* different from their component parts' in Chapter 17. At this stage of the argument we have to bear in mind that relevant knowledge is historically, and therefore socially, defined.

[T]he view of the universe and a particular kind of society holding this view are closely interdependent. They are a single system, neither can exist without the other.

(Douglas 1980: 289)

The environment is culturally defined through the way in which people socially, and therefore politically interact to survive through a production process, with particular technical properties.

'Environment' is a social category because 'environment' does not exist without somebody defining it as an 'environment' of something. If the defining subject disappears, the environment disappears. Only social human beings are capable of such definitions. Consequently, environmental research is intrinsically laden with social considerations, and criteria used in assessing the quality of the environment are based on some point of view.

(Haila and Levins 1992: 113)

But having situated the concept of the environment in its historical and social context, there is still a common material reality within which people survive, and which sets the parameters to human social behaviour. And the particular way in which the environment is socially experienced colours people's interpretation of this common, material, ecological reality, biasing the understanding of the environment, and suggesting policy initiatives that favour particular economic and social interests. It is the purpose of scientific inquiry to identify and define these material parameters, so that the bias of understanding can be highlighted, and people can consciously chart a course for the utilization of the environment in the process of social development, maximizing the potential for human creativity (see Chapter 13).

[T]he social and cultural dimension of environmental problems must be understood if we are to achieve understanding and practical action.

(Irwin 1995: 37)

Development processes are inherently indeterminate and unpredictable, and hence 'history becomes of paramount importance' to purposively intervene and direct the pace and direction of social change and evolution.

With regard to the exploitation and control of the natural environment, and human 'victories' *over* nature:

> for each such victory, nature takes its revenge on us. Each victory, it is true, in the first place brings about results we expected, but in the second and third places it has quite different unforeseen effects which only too often cancel the first ... At every step we are reminded that we ... belong to nature, and exist in its midst, and that our mastery over it consists in the fact that we have the advantage over the other creatures of being able to learn its laws and apply them correctly.
>
> (Engels 1940: 291–2)

The ability to 'learn its laws and apply them correctly' will reflect the culture of knowledge in society: the hegemonic ideology of science, through which social experience is explained and understood. A cultural hegemony reflects and legitimates the interests of the ruling class in society, justifying the social status quo (see Chapters 17 and 18). For sociocentric theorists it is precisely this cultural bias in understanding that underlies environmental problems. Solutions call for a change in social customs, practices and relations, which will alter in response to people's struggling to fulfil their social potentials in the face of powerful vested interests (see Chapter 13 and Part IV).

Problems are defined with respect to the dominant cultural values of society.

> [A]s with all preceding ruling classes, the bourgeoisie has always had the ability to generalise its own particular class view of the world as being a world view, a universal truth.
>
> (Burgess 1978, quoted Pepper 1993: 95)

Environmental crises and commodity exchange

The environmental crisis is in a real sense a crisis of how we socially interact through institutions to utilize the natural environment. The dominant mode of social organization in the world today is the capitalist mode of production, and the ruling class the 'bourgeoisie' (see Chapter 13). In the capitalist mode of production people socially interact, and political power is exercised *through* commodity exchange (see Chapter 5).

> We abuse the land because we regard it as a commodity belonging to us. When we see the land as a community, to which we belong, we may begin to use it with love and respect.
>
> (Leopold 1968, quoted Harvey 1993: 3)

Where social life is conducted through commodity exchange, the objective being to realize a profit on relations of exchange, then there are certain logical parameters to rational, individual, social behaviour (addressed in Chapter 5). And with a globalized world economy, individuals' particular, culturally specific behaviour obeys a universal social imperative: the purpose of economic activity is to maximize profits, and only incidentally to satisfy social need. In the current context of economic policy intended to ameliorate the trend towards ever more severe economic crises, reflecting a trend for the profitability of relations of commodity exchange to decline (see Chapter 5), the International Monetary Fund (IMF), the World Bank (WB), and the General Agreement on Tariffs and Trade (GATT) are attempting to avoid a rerun of the international economic crisis of the 1930s.

The 'crash' of the Wall Street Stock Exchange on 29 October 1929 was a consequence of the tendency for the rate of profit to fall (see Chapter 5), in conjunction with government's economic policy intended to mitigate the effects of the world crisis of the capitalist mode of production on their national electorate. The national interest of the bourgeoisie was asserted over the (international) class interest of the capitalist ruling class. Governments attempted to stave off the drift towards economic recession – with the falling rate of profit more and more enterprises went bankrupt, ceased to trade, with the consequent spiralling of unemployment – by in effect subsidizing economies through the devaluation of national currencies. However, in as far as any particular economy could ameliorate declining trade, and hence falling profitability, by an arbitrary devaluation that increased international demand, this only served to exacerbate the problems felt in other economies. So when the *actual* value of production, and hence the value of enterprises, was finally registered in stock markets around the world, the fall in the value of stocks and shares was precipitate and unexpected. Hence the October 1929 stock market crash.

The social effects of this economic crisis reverberated for another fifteen years, culminating in the Second World War. To this end, in the era of the globalized economy, characterized by capital that is internationally mobile in search of profitable opportunities to produce and trade, economic policy must not be allowed to be dictated by a nationalist, social agenda. The managed depression of the 1980s and 1990s through the structural adjustment programmes sponsored by the World Bank and the IMF (see Chapter 11) has seen the trend towards unemployment and the bankruptcy of enterprises *gradually* deteriorate (rather than a crash). In this context, the International Labour Organization estimates that there are at least 1 billion people in the world who are unemployed. But the resultant misery is none the less if the decline is gradual, and the immediate future portends ever greater protest, born of frustration, as people try to fulfil their potentials. And ever greater repression is likely as the political powers within the capitalist mode of production are unable to satisfy people's expectations.

They [IMF, WB, GATT] have encouraged and coerced third world countries to lower their import quotas and tariffs, which protected their
fledgling industries, to devalue their currencies, making their exports
cheaper and Western imports dearer, to cut welfare expenditure . . . in
[an] attempt to 'recolonise' the third world . . .

(Pepper 1993: 104)

The spiral towards social disintegration and barbarism will get ever deeper,
until there is an awareness and a consciousness of a shared class interest
amongst the disadvantaged, an interest that can be prosecuted through political mobilization and organization.

As an illustration of how the world market structures the production of
nation states, generating frustration, economic and political struggle, and
environmental degradation, see Jenny Pearce's book *Promised Land* (Pearce
1986). In *Promised Land* the political economy of El Salvador is analysed
from the sixteenth century to the present day, with crops intended for the
world market – indigo in the seventeenth century, coffee in the nineteenth
century and cotton in the twentieth century – successively marginalizing
peasants onto ever more unproductive land, inevitably generating land degradation, poverty and political unrest.

Social relations, then, structure the form of economic transactions through
which people interact with the natural environment, which feeds back and
sustains those social relations of production in a dialectical process of
change. Driven by people's frustration at being denied the social opportunity to fulfil themselves, society and the environment co-evolve towards a
mode of existence characterized by enhanced human potentials (see Chapter
13). People are *not* passive beings moulded by biology (egocentrism), or
the environment (ecocentrism); they *actively* create their own conditions
of existence.

[T]here is not [an] environment in some independent and abstract
sense. Just as there is no organism without an environment, there is no
environment without an organism. Organisms do not experience environments they create them.

(Lewontin 1991: 109)

Humans do not passively mould themselves to environmental forces beyond
their control, in a Darwinian process of natural selection.

Organizms within their individual lifetimes, and in the course of their
evolution as a species, do not *adapt* to environments: they *construct*
them. They are not simply *objects* of the laws of nature, altering themselves to the inevitable, but active *subjects* transforming nature according
to its laws.

(Richard Lewontin, quoted Harvey 1993: 88, emphasis added)

Human organisms transform nature through social relations, and the social relations of commodity exchange, implying contradictory class interests in social existence (see Chapters 5 and 13), are the particular focus of interest of sociocentric theorists. These relations, which generate such environmental phenomena as urban development (Smith 1990), are ideologically legitimated and sustained, and, consequently, any attempt to contain deleterious environmental processes has ultimately to challenge the extant ideological hegemony. In this regard, the liberation ecologists (Peet and Watts 1996a) consider the strategy of the political ecologists to be extremely limited. Liberation ecology is intended for

> a practical political engagement with new movements, organizations, and institutions of civil society challenging conventional notions of development, politics, democracy, and sustainability . . .
>
> (Peet and Watts 1996b: 3)

Problems are not the consequence of poor management, inappropriate (or lack of) technology, inadequate expertise, the pressure of population growth, etc., but of political-economic constraints on individuals' social behaviour. And the theorist has to take sides, decide which interest to support. The contradictions are not logical, as the political ecologists maintain, but *political.* Consequently problems are not amenable merely to rational solution; environmental problems and issues are more than logical problems, to be solved by consultants or academics. There may well be pragmatic compromise between competing understandings, attitudes or points of view, which might suggest a logical outcome to the conflict; but such compromises reflect the balance of power in society and not the rational resolution of intellectual problems.

The theorist has to become the activist, and not the disinterested expert. The conflict is essentially one of competing rationalities intended to achieve different ends, reflecting differential experience which generates contradictory interests, which are not to be rationally solved by appropriate management: for instance, the pressures of production transmitted to the environment, reflecting social relations based upon commodity exchange, which compel land managers to make excessive demands on the environment.

> . . . facts of degradation are contested and . . . there will always be multiple perceptions (and explanations) . . .
>
> (Peet and Watts 1996b: 6)

Post-structuralist, postmodernist or discourse theorists, noting different perceptions of the same reality, having no basis for comparison, conclude that alternative points of view are equally valid. Such ideologies offer no guidance as to the interpretation of a shared reality which is implicit in alternative theoretical representations of experience. And therefore they have nothing to say as to what *should* be the policy prescriptions for shared environmental problems. This is a theme to which we shall return in different contexts in Part IV of this book. But at this stage of the argument we have to bear in mind the social construction of theoretical understandings of experience.

> Like other ideologies environmentalism is socially constructed . . .
>
> (Buttel 1992: 15)

The social construction of environmental problems

It is not that exogenous natural forces act on producers so that they despoil the environment – as Blaikie and Brookfield argue with regard to peasants and land degradation (Blaikie and Brookfield 1987: 70). Producers are part of a complex web of competing, contradictory social relations. People with distinct political interests rationally behave in conflicting ways, and those who suffer are the disadvantaged and the powerless. Hence the concepts of power and resistance must be coherently integrated into environmental analyses. Everyday resistance in the community – trades union movements, protest through party political institutions with regard to resource control and allocation, neighbourhood action committees, etc., with the focus of property rights, workers' employment rights and obligations, welfare rights in the community (for instance education and health) – has to be addressed. It is not a question of whether or not our behaviour accords with nature; it is more a question of 'Do we like what we have produced?' – shopping malls, ten-lane motorways, toxic waste, nuclear power, slums and shanty towns, unemployment and economic recession, water and air pollution . . .

> All critical examinations of the relation to nature are simultaneously critical examinations of society.
>
> (Harvey 1993: 39)

Specific social relations structure the way human labour interacts with, and transforms, the natural environment.

> An ecological history begins by assuming a dynamic and changing relationship between environment and culture, one as apt to produce contradictions as continuities. Moreover, it assumes that the interactions of the two are dialectical. Environment may initially shape the range of choices available to a people at a given moment but then culture reshapes environment responding to those choices. The reshaped environment presents a new set of possibilities for cultural reproduction thus setting up a new cycle of mutual determination. Changes in the way people create and re-create their livelihood must be analysed in terms of changes not only in their *social relations* but in their *ecological* ones as well.
>
> (Cronon 1983: 13–14, quoted Harvey 1993: 27, emphasis added)

Nature has no power or will, and does not exact revenge when we trans-
gress her laws. In as far as humans, socially organizing themselves, *do* degrade
the natural environment it is our *social* 'fault', and not outside our control;
though of course the powerless who invariably suffer the full effect of environmental degradation are politically least able to alter the course of events.
To see nature as capable of human emotions, to take 'revenge', is to fetishize
nature – to invest the environment with a human will and emotion. Such a
fetishism disempowers people from controlling their own lives; fatalistically
nature is empowered with an independent will beyond our control. Invariably
the effect of such myths is to protect and preserve an iniquitous social order.
For Marx such attitudes are

> characteristic of an alienated social order – an order in which facts about
> humans and human relations are falsely projected onto the world of
> objects.
>
> (Vogel 1988: 379)

People are essentially creative, social individuals; their creative potentials reflect
their social experience. When people are constrained from fulfilling themselves through social obstacles beyond their control – for instance they might
become unemployed – then the resultant frustration at not being able to
realize their potentials can be either destructive or constructive. If the anger
and frustration are internalized, and turn inward on individuals – either people
blame themselves for their fate, seeing themselves as inadequate or incompetent, or the frustration manifests itself through anti-social behaviour,
violence and abuse – then the social frustration of individual activity only
destroys the individual. If, however, people are aware of others enduring
similar obstacles to achieving their potentials, and these people recognize a
shared interest, which they prosecute through mobilization and political organization, then potentially these frustrations can lead to positive changes in
society. Society evolves towards better fulfilling people's creative potentials:
development (see Chapter 13).

Marx's intellectual project was to analyse in detail the capitalist mode of
production, in order to identify the social process that leads to inequality,
denying the powerless the opportunity to fulfil their creative potentials. And
the dynamic of this process is commodity exchange, the defining characteristic of the capitalist mode of production being that labour power *itself* is a
commodity; necessarily people's potentials are fulfilled according to the
vagaries of market exchange, and the resulting frustrations allow for the possibility of fundamental social change (see Chapter 5).

Individuals' frustrations are conceptualized in terms of individuals being
'alienated' from their nature as creative, social beings, denying the fulfilment
of 'historically created possibilities' (Petrovic 1983: 13).

Given the thrust of Marx's work, there is little explicit reference to *the*
'environment', just as there is little discussion of 'socialism': the intellectual

project was to identify the important social contradictions, those with the potential to create frustrations the resolution of which would fundamentally challenge the social order (in nineteenth-century Britain and Europe). The issue was not what would be the social frustrations a hundred years hence, nor the detail of what the alternative might be (see Chapter 13). Ted Benton (1989) argues that Marx's emphasis on exploitation and commodity exchange led to his ignoring the ecological limits to economic and social reproduction – naturally given limits cannot be treated as mere means of production. Such an assessment reflects a lack of understanding of the dialectic of social change (treated in detail in Chapter 13). But the failure to understand Marx's method is common in academic circles in general, and environmental debates in particular, so that Marxism 'appears as a stranger in ecological circles' (Kovel 1995: 31).

In nineteenth-century Europe, the main obstacles to people's fulfilling their creative potentials were economic exploitation and social deprivation, rather than obviously ecological issues. And Martinez-Aller (1990) is wrong when he argues that the lack of an ecological theory in Marx is a result of 'Marx's view of history . . . [which] envisages *unlimited* development of the productive forces under socialism' (Pepper 1993: 60, emphasis in original). Similarly, Michael Redclift thinks that Marx overemphasized commodity production to the neglect of the 'natural processes of social reproduction', and he supports Andrew Sayer's (1983) objection to Marxism's emphasis on how 'social forms, including economics, mediates nature' (Pepper 1993: 61). We shall return to a consideration of the contribution of a Marxist method to an understanding of social development in general and the environment in particular in Chapters 13 and 17.

However, at this stage of the argument, it is important to bear in mind that Marx was trying to explain how changing social constraints frustrate individuals from fulfilling their changing creative potentials, and how this might lead to social development – to a society more in accord with human nature (see Chapter 13). 'Marxists offer a dialectical view of the society–nature relationship' (Pepper 1993: 107), in which, through the transformation of nature, humans are themselves 'transformed . . . nature is definable as the materials and forces of the environment that create man and are in turn created by man' (Parsons 1977: xi).

> [T]hrough learning how to farm nature's products, we changed ourselves from nomadic hunter/gatherers to sedentary people. Through learning how to manufacture things we changed ourselves to an industrial society. Agricultural and industrial societies, and the individuals in them, are *qualitatively* different from each other, and from what preceded them . . . As our ability to use resources has grown we have developed new needs: housing, energy, telecommunication. As we have changed our power to do things (e.g. via transport, computers) we have changed the things we want and need . . . the society–nature dialectic develops

> to produce ... new kinds of individual ... Creating the machine ...
> has brought new moral and cultural values to ... society ...
>
> (Pepper 1993: 112, 113, emphasis in original)

Marxist theory is sensitive to how people fulfil their potentials through socially interacting with ecology, in order to reproduce themselves and survive. And in the contemporary world the capitalist mode of production defines the rationality of social interaction (see Chapter 5), but 'capitalism ... dehumanizes man and perverts the natural world' (Vaillancourt 1992: 34). For ecocentric theorists, the social system has to adapt to the ecological system – that is, these are two systems that function independently, but are interdependent (though not dialectically related, they do not co-evolve). People have to learn to live by the ecological limits to social existence. For Marx ecology is not a system, but an integral part of a changing process of human existence, through which individuals fulfil their changing potentials: *development*.

Consequently, for Marxist theorists, while environmental degradation might be a *consequence* of industrialization and/or poverty, it is not *caused* by industrial development or inequality. Within the capitalist mode of production, where the intellectual parameters of social behaviour are set by commodity exchange, industrial developers quite rationally (in their terms) despoil the environment, and the normal functioning of capitalist society creates inequality. Capital accumulation, as a consequence of commodity exchange, demands ever higher levels of efficiency from labour in the production process. The consequence is: processes of urbanization; industrial development; a competitive economy where the security of individuals to fulfil their potentials is ever more tenuous; which in turn suggests the breakdown of social cohesion and rising levels of political coercion and repression to maintain social order; etc. With regard to the environment:

> Capitalist production develops the technology and the combination of social production processes, at the same time exhausting the two sources of all wealth: land and labour.
>
> (Marx 1974, quoted Leff 1992: 110)

Understanding the dialectic of social change is the key to reintegrating Marxist theory and ecological thought. Although this is addressed in different contexts in Chapters 5, 13 and 17, the important insight for an understanding of the environment is: if humans are considered to be creative, social individuals, whose potentials evolve with social experience, then the fundamental dialectic of social change is between individuals' potentials and social opportunity, opportunities that have an ecological dimension. Within the capitalist mode of production the dynamic of capital accumulation through commodity exchange recasts the contradiction between individual fulfilment and social opportunity, as one between the *forces* and *relations* of production (see Chapter 5). That is, the contradiction between the forces and relations of production is the way in which the deeper, 'essential' contradiction within the individual *appears* in social activity. And within the capitalist mode of production people relate through commodity exchange.

> [A]ll science would be superfluous if the outward *appearance* and *essence* of things directly coincided.
>
> (Marx 1972a: 817, emphasis added)

The dialectic, the essential contradiction that explains social change, is *within* people. To confuse the appearance (the contradiction between the forces and relations of production in commodity exchange) with the essence (the contradiction between human nature and social existence) is to fetishize the commodity: to invest *what* we exchange with the will and motivation of *who* is party to the exchange. Hence it appears as if the forces and relations of production cause economic crises. They do not; *people* do. And for some Marxist theorists environmental degradation is analysed, and fetishized, at the level of commodity exchange. For instance, James O'Connor argues that the first contradiction of the capitalist mode of production is that between the forces and relations of production, to which has to be added a second: the environmental limits to capital accumulation. And it is this second contradiction that provides the basis for an 'ecological Marxism'.

> [T]he point of departure of an 'ecological marxist' theory of economic crisis and transformation to socialism is the contradiction between capitalist production relations (and productive forces) and the *conditions* of capitalist production, or 'capitalist relations and forces of social reproduction'.
>
> (O'Connor 1988: 16, emphasis added)

O'Connor's point is that the natural, environmental conditions of production, which along with labour power and the means of production go to produce commodities, should not in themselves be treated *as* commodities. They are not reproducible as commodities, and in as far as they are not reproducible, capitalist production is environmentally unsustainable: natural commodities are 'not produced and reproduced capitalistically' (O'Connor 1988: 23). O'Connor is fetishizing commodities: for Marxist theory the commodity is not a thing but a relationship: it is an *explanation* of social behaviour. Hence, in as far as environmental resources are part of relations of exchange *between* people, they *are* commodities, whether or not they are reproducible. And it is the failure of relations of commodity exchange to fulfil the potential of human nature that explains any possible socialist transformation, not the contradiction between the forces and relations of production, or the non-reproduction of 'natural' commodities (see Chapter 13). It is not that there are *two* contradictions, but the fundamental contradiction that explains social change appears in different ways in different contexts of social behaviour. '[Are] there . . . two separate contradictions of capitalism, or, alternatively, two forms of one contradiction[?]' (Lebowitz 1992: 92).

By mistakenly fetishizing market categories and environmental concepts, in effect, the contradiction that generates social change lies outside individuals' motivation; this inadvertently opens the possibility of authoritarian (if benign) politics of social change towards socialism. The emphasis is moved away from socialist transitions as greater individual self-awareness, self-consciousness and the fulfilment of individuals' potentials, to the control of the forces and relations of production (see Cole and Yaxley 1991, and Chapters 13 and 17).

Where social change is understood as a product of individuals' social frustration in the fulfilment of their creative potentials, emphasis is given to people being aware of shared, class interests, along with having the skills and confidence to collectively mobilize and prosecute a class-based political agenda. Hence *ideology*, the thought processes and intuitions through which people perceive and understand their lives, is fundamental to achieving meaningful social change and development.

The role of ideology

In this context Barry Commoner (1990) argues that 'meaningful environmental improvement requires the *proper* choice of techniques and systems of production' (Commoner 1990: 154, emphasis added), but within the capitalist mode of production, based upon competitive market exchange, that choice 'is in private, not public hands' (Commoner 1990: 154). Decisions are taken with regard to short-term economic considerations of concern to stockholders and the return on their investment, rather than with regard to the long-term social effects of investment decisions. This raises questions about the social control of corporate power, and mounting an ideological challenge to the notion that market exchange is the rational way to organize social activity. However, it is clear from the environmental problems in the former Soviet Union – for example, the nuclear contamination emanating from the nuclear submarine bases in northern Russia, the effect of the explosion at the Chernobil nuclear power station, and the water and air pollution from Soviet industry – that the state is not necessarily any better than the market at ensuring our ecological well-being (Feshbach and Friendly 1992). For Commoner the problem is technological.

> It is an unassailable if ironic fact that the economic development of the Soviet Union and its socialist neighbours after World War II has been based on the major new production technologies developed in the United States and other capitalist countries, where their design was guided by a capitalist motive: short-term profit maximization . . .
>
> (Commoner 1990: 159)

He concludes that there is no institutional example on which to base the social governance of technical decisions, and hence on how to include public discussion in the solution of environmental crises.

> It seems evident, therefore, that the responsibility for introducing social concerns such as environmental quality into corporate production deci-sions lies with society itself.
>
> (Commoner 1990: 163)

But how?

For Commoner, it is where there is a social demand for democracy – in East Germany people deciding to pull down the Berlin Wall; in Czechoslovakia people rallying to bring down the government imposed by the Soviet Union after the 1968 invasion; in Romania people deposing an autocratic ruler and instituting democratic reforms; in Poland the autonomous union Solidarity ending one-party rule; in the Soviet Union the reforms of *perestroika* and *glasnost* inspired by Mikhail Gorbachev leading to the end of one-party rule – that change takes place. And in the West, the social movements for civil rights and environmental protection, and against gender, racial and religious discrimination, are similarly an expression of people's need to participate in social organization.

> . . . in the march on Selma Alabama; the students occupying Tiananmen Square; the crowds pouring into the Potsdam Platz and Wenceslas Square; the huge demonstration for nuclear disarmament in Aldermaston, England, or Washington DC; the strike at the Lenin Shipyards in Gdansk; or for that matter, decisions made by a few men in the Kremlin – confirms the existence of a common desire for everyday democracy that is deeply felt, powerful, and eventually undeniable.
>
> (Commoner 1990: 165)

The power of corporate managers and state planners has to be subject to democratic control. In the west, 'green' political parties have not galvanized the collective mind of the electorate, in part because of the challenge that green poli-cies pose to industrial development, and the threat to employment of the green-ing of industrial production. Such a green political agenda is particularly a problem in less developed economies, where there is little potential to invest in ecologically sound production processes, and the need to generate employment is the greatest. In the 1992 US presidential campaign George Bush, running for reelection, criticized environmentalists for 'putting Americans out of work'.

What is the solution?

> . . . a democracy that encompasses not only personal and political freedom, but the germinal decision that determines how we and the planet will live.
>
> (Commoner 1990: 176)

That is, people of the world have to understand that survival is contingent on peace between people, and with the environment. But beyond the admonition of the need for an ideological sea-change in the way people understand social existence, there is little in Barry Commoner's thinking and writing to suggest how this might be achieved.

Richard Levins and Richard Lewontin, however, offer a deeper understanding of the process of realizing such an ideological, and ultimately political, change in individuals' thinking and in social organization. It is not a question of society coming to terms with a pre-existing, extant environment, a position that implies that there is an ecological, evolutionary logic to which humans have to adapt. Humans and the environment are part of the *same* reality, which 'is a contingent structure in reciprocal interaction with its own parts' (Levins and Lewontin 1985: 136).

> [E]nvironments are as much the product of organisms as organisms are of environments . . . There is no organism without an environment, but there is no environment without an organism.'
>
> (Lewontin and Levins 1997a: 96)

Environments are not imposed on people externally, but through evolutionary processes they are socially created. In turn the environment provides the preconditions for future generations. Humans are both the consumers and producers of the resources necessary for their own existence. The implication is that humans can consciously intervene to mould the environment through changed social activity and social experience. But the dialectical relation of mutual determination – the co-evolution – of society and the environment implies a degree of indeterminacy: 'every population species and community, indeed the whole damned biosphere, is constantly changing in what appears to be unpredictable ways' (Lewontin and Levins 1996: 96). This does not imply that there is no causality, only that behaviour is in part a consequence of subjective understandings about life reflecting differential experience. Therefore ideas about what the future *might* be is conditioned by our perceptions of what the world *is*, or has been.

> While the consciousness of an individual is not determined by his or her class position but is influenced by idiosyncratic factors that appear as random, those random factors operate within a domain and with probabilities that are constrained and directed by social forces.
>
> (Lewontin and Levins 1997b: 68)

Hence a study of social evolution and the social potential of individuals' activity might suggest possible social contradictions and class struggle, and a strategy for addressing environmental issues.

One cannot make a sensible environmental policy with the slogan 'Save the Environment', because 'the' environment does not exist and second because every species, not only human species, is at every moment constructing and destroying the world it inhabits.

(Lewontin and Levins 1997a: 98)

We have to ideologically understand the social dynamic of human existence, recognizing conflicting class interests: we have to be aware of the intellectual parameters of our world (see Part IV). That is, we conceive of the world through culturally specific concepts that reflect our experience, and which shape our perceptions of that world in an image that typically, culturally, justifies an iniquitous social status quo.

> . . . science . . . is a social institution completely integrated into and influenced by the structure of all our other social institutions . . . Scientists . . . view nature through a lens that has been moulded by their social experience.
>
> (Lewontin 1991: 3)

In Part IV we shall return to the issue of just how we can begin to construct an image of the world that attempts to traverse the ideological limits of scientific inquiry, and which raises important questions of the role of the intellectual in the production of knowledge. *All* proposals to alter patterns of social organization and behaviour aimed at resolving environmental problems adversely affect some people, who may or may not be in a position to exert political power to protect their interests. And in class-based societies, where people have to compete for their livelihoods, often those people who would benefit most from policies intended to improve the environment are also those people most economically threatened by such initiatives.

> It is no use saying to South Wales [coal] miners that all around them is an ecological disaster. They already know. They live in it. They have lived in it for generations. They carry it in their lungs . . . in all too many environmental cases . . . there is a middle-class environmental group protesting against the damage and there's a trade union group supporting the coming of work.
>
> (Williams 1995: 51)

In the process of environmental, and hence social, change, people must intuitively understand their individual experience as part of a changing process of social relations. A subjective awareness of shared material interests that can provide the rationale for purposeful, collective, political action to change society will in part reflect the experience of *class-consciousness*. As we shall see in Chapters 13 and 17, the concept of 'class' does not describe lifestyles, but *explains* interests: and class-consciousness is not an intellectual attitude that people possess, it is something people experience, and in the process of the integration of practice and theory, *praxis* alters consciousness.

It is not a question of the correct environmental principles structuring social life; nor is it an issue of defining politics in ecological terms, of social organization and behaviour being within the confines of natural laws. Social life is riven by conflicting political interests, which define the limits to individuals' fulfilling their creative potentials, and which cannot be resolved through rational debate. People need to be conscious of where their interests lie; with whom they share these interests; what the scope might be for political mobilization and organization; and the strength of competing interests to be able to prosecute these interests through political struggle.

> The democratic task of science is to make the arcane obvious and that means educating the intuition, which we conceive of as an integration of experience, knowledge and feelings rather than as a mythical antithesis to reason . . .
>
> (Levins and Lewontin 1994: 36)

Abstract class interests become intuitive in a process of changing cultural hegemony, and class-consciousness characterizes social behaviour the experience of which changes perceptions of 'the' environment (see Chapter 13).

Forms of social relations culturally mediate humans with physical conditions:

> [N]ew environmental impacts are created. A severe winter in an urban environment does not produce frostbite, but hunger, when the poor divert resources from food to fuel. Racism becomes an environmental factor affecting adrenals and other organs in ways that tigers or venomous snakes did in earlier historical epochs. The conditions under which labor power is sold in a capitalist labor market act on the individuals' glucose cycle as the pattern of excretion and rest depends more on the employer's economic decisions than on the worker's self-perception of metabolic flux. Human ecology is not the relation of our species in general with the rest of nature, but rather the relations of different societies, and the classes, genders, ages, grades, and ethnicities maintained by those social structures. Thus, it is not too far-fetched to speak of the pancreas under capitalism or the proletarian lung.
>
> (Lewontin and Levins 1997c: 90–1)

On the analysis of environmental conflicts as social and political struggles see Fitton *et al.* (1994), Faber (1998), and Levenstein and Wooding (1998).

Certainly we need expertise to identify ecological processes that threaten social life, but with regard to solving the problem, this is only half of the analysis. The other half is engaging with the population to raise people's awareness of the social processes by which the environment is degraded: social processes that are relevant to people's experience, and not the intellectual, theoretical concerns of the 'expert'. It is not CFCs that threaten the ozone layer, but an industrial way of life for which these gases are expedient in both production and consumption, and in as far as this way of life can be changed then the threat to human life from these gases may be alleviated. Any change in the way we socially interact with the physical environment will adversely affect someone's lifestyle or livelihood somewhere. Changes in the way we use or allocate resources have a necessary political dimension.

> If you only pick up the physical appearances, you are likely to miss all the central social and economic questions, which is where ecological thinking and social thinking necessarily converge.
> (Raymond Williams, quoted Haila and Levins 1992: 57)

When competitive relations of commodity exchange characterize social life, contradictory class relations impede political decision making intended to develop society towards the fulfilment of people's creative potentials. This is the process of socialist development (see Chapter 13).

Part III
Development

Development studies

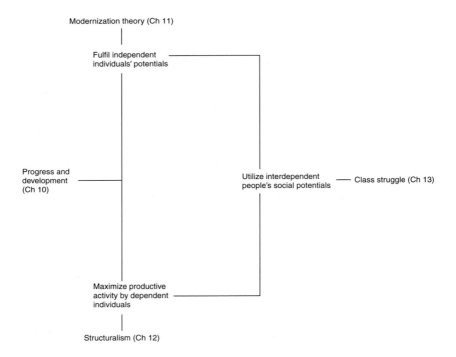

Modernization theory (Ch 11)

Fulfil independent
individuals' potentials

Progress and
development
(Ch 10)

Utilize interdependent
people's social potentials — Class struggle (Ch 13)

Maximize productive
activity by dependent
individuals

Structuralism (Ch 12)

10 Development

Development

Everybody is in favour of 'development': everybody wants to see 'progress'. And everybody agrees that development implies social change. For societies to change, people have to change their behaviour, adapting their expectations and ambitions towards goals that advance individual and social life: that is, objectives that either improve people's standard of living, or make their lifestyles more desirable.

While there is broad agreement over *what* development is, there is little agreement as to *how* development is achieved, or *why* change should mean progress.

In recent times debates about development have revolved around the idea of development being *sustainable*. How can processes of social change be directed towards *progress*, and more particularly, how can such social advance be maintained into the foreseeable future? Such was the agenda of the United Nations Conference of Environment and Development (the Earth Summit) of June 1992, addressed in Chapter 6. In the closing years of the twentieth century concern about the international environmental consequences of national development has been compounded, and in some cases eclipsed, by ubiquitous *intra*-national conflicts, which all too often erupt into civil wars.

Increasingly in the 1990s conflicts are recognized as being a consequence of disputes *within* nation states over the legitimacy of power, in which in the post-1945 period the 'super-powers' (the USA and the Soviet Union) intervened to try to ensure that the victors of such conflicts were the 'allies' of one or the other, in the communism versus capitalism, East–West Cold War, conflict. It is estimated that in 1992 alone over 500,000 people died in conflicts in Afghanistan, Armenia, Bosnia, Cambodia, Croatia, Georgia, Iraq, Palestine, Serbia, Somalia, Tadjikistan, Turkey, Albania, Zaire, Angola, El Salvador, Guatemala, Mozambique, Bangladesh, Burma, Colombia, Indonesia, Liberia, Nigeria, Peru, Rwanda, Sudan, and Sri Lanka (reported in the *Guardian* 31 December 1992, and see also Brogan 1992 for a more comprehensive analysis). And since the Second World War there have been over 130 such conflicts in which millions have died, been disabled, injured, or displaced, becoming refugees.

What does this tell us about development, and development policy?

Clearly people's lives and well-being are fundamentally affected by such cataclysmic processes, and no understanding of development and progress can ignore such catastrophes: they are too pervasive.

While development theorists are still concerned with deriving policies and strategies that would make development sustainable, to meet 'the needs of the present without compromising the ability of future generations to meet their . . . needs' (World Commission on Environment and Development 1987: 8), and in particular how inequality, disadvantage and poverty lead to the despoliation of the environment, and ultimately unsustainable development, it is clear that sustainability also has a political dimension. And in this regard, contemporary development debates increasingly address the concepts of *empowerment*, *participation*, and *globalization*.

But academics, development workers, experts and development consultants working within the intellectual parameters of different theoretical perspectives understand the same experience of social change quite differently, implying different policy strategies to advance progress. 'Progressive' social change improves people's lives. And what such change should be, and how it can be effected, reflects an understanding of why there is inequality, disadvantage and conflict in the first place. This implies a conception of why people behave in particular ways: what is human nature?

The significance of beliefs in human nature for social inquiry will be addressed in Chapter 18: and the science of asking 'what', 'how' and 'why' questions will be discussed in Part IV (Chapters 14–17). This section of the book (Chapters 10–13) is intended to explain the differences between perspectives on development.

Independent individuals

For some theorists, individuals' characteristics and talents, motivation, etc. are believed to be *innate*: a genetic 'endowment'. All that development theorists and development policy can hope to achieve is to create an environment in which individuals are able to fulfil the potentials inherent in their biological inheritance. Progress is the process of creating a culture, and the associated institutions, that will not only allow people to achieve what they are personally capable of, but also facilitate an awareness of their individual interests: a society based on individual responsibility.

Such a process is conceived of as the *modernization* of society (see Chapter 11). Fundamental to modernization theory is the institution of a market economy, allowing individuals to behave according to their subjective preferences, reflecting their endowment of tastes and talents (see Chapter 3). The fostering of functioning markets necessitates a range of ancillary political, social, religious, legal and cultural institutions, and it is the development project of modernization theory to create such institutions, a project that leads to progress only in as far as people accept individual responsibility for their lives: the modernization of individuals' *minds*.

Economic development requires *modernization of the mind*. It requires revision of the attitudes, modes of conduct and institutions adverse to material progress . . . economic achievement and progress largely depend upon human aptitudes and attitudes . . .

(Bauer 1976: 84, 41, emphasis added)

In this regard the developed market economies of the world, North America, Western Europe and Australasia, have enjoyed such a modern, cultural and institutional context to individual existence for the past two or three hundred years, which for modernization theorists explains the relative affluence of these societies. Clearly, within these societies there is still relative poverty, but such inequality tends to reflect individuals' inadequacies, and does not fundamentally call into question development as modernization.

Society is as equal as it can be, *given the natural inequalities between people*. According to this view, the political and social revolutions [in Europe and America] of the eighteenth and nineteenth centuries destroyed artificial hierarchies and allowed the *natural* differences in ability to manifest themselves.

(Lewontin 1982: 89, emphasis added)

Poverty is either caused by a lack of modern social institutions; or the poor themselves are to blame. The former can be addressed by modernization theorists, who are able to identify the institutional constraints on people being able to act *naturally*: to act instinctively according to their *human nature*.

However, some modernization theorists believe that even if underdeveloped countries could avail themselves of such 'modern' institutions, the majority of people in these societies would *still* be poor. It is argued that since humans are sentient beings, behaviour has inevitable psychological antecedents, and although at present, with regard to human motivation and psychology

no scientific proof is possible . . . psychology poses questions to which biochemistry might find answers, which may be further explained by genetic coding . . . Assuming that certain mentalities are not conducive to socioeconomic development, the question of whether or not they are *innate* . . . becomes important.

(May 1981: 247, 251, emphasis added)

That is, the question of the biological endowment of the facility for social progress and development is assumed, *believed* to be important – a recasting of the sociological theory of Max Weber.

When we find again and again that, even in departments of life apparently mutually independent, certain types of rationalization have developed in the Occident, and only there, it would be natural to suspect that the most important reason lay in differences of heredity. The author admits that he is inclined to think the importance of heredity very great.

(Weber 1974: 30)

Rationalization and motivation become a question of 'comparative *racial* neurology and psychology' (Weber 1974: 30, emphasis added). For Brian May, Weber's phenomenon of Occidental prosperity and economic dynamism is the antithesis of Third World poverty and stagnation, and he asks the question: Can the Western trait of being 'rational' and 'calculating' be attributed to a differential biochemistry in the brain of different races, the way in which the 'reasonable' left half of the brain interacts with the 'aesthetic' right half of the brain? And because 'genetically determined blood characteristics [of] the human population may be classed in five main races: African, American, Asian, European and Pacific', this suggests that there are other genetic variations that can be used to explain why 'irrational pressures in the Third World are increasing' (May 1981: 254–5), evidence of which in the 1960s and 1970s was the rise to power of Idi Amin in Uganda, Pol Pot in Cambodia, and Emperor Bokassa in the Central African Republic: all classed as tyrants. Either way, May thinks it was 'One of the great disservices to mankind performed by the Nazis . . . to reduce the question of race to a sordid emotive issue . . .' (May 1981: 254), an issue that he believes is fundamental to examining Darwin's theory of natural selection and genetic drift, the survival of the fittest, the prosperity of Western societies, and the understanding of development.

However, not all modernization theorists explicitly correlate development with race, though this is a possible implication of the logic of the modernization 'schools' of thought, such as: stage theory, dualism, structural adjustment, and new institutional economics, all of which are addressed in detail in Chapter 11.

Dependent individuals

As we saw in Chapter 3, economists working with the cost-of-production theory of value believe that individuals cannot be considered to be independent; they are *not* free to choose, living as they do in a society in which they have to learn to cooperate to survive, rather than compete to further their own advantage. Such an approach has parallels in development theory. Resources and productive activity have to be managed in the social interest. And now poverty and disadvantage are not merely a consequence of a lack of individual freedom and/or individuals' personal inadequacies. Certainly some individuals may appear to be incorrigibly reticent about participating in productive activity, but such behaviour is not judged to be a reflection of an innate, genetic endowment. Not only are individuals technically, and hence socially, dependent for their survival, but people learn to adapt to the needs of society: individuals are *socialized* to accept the mores and customs of society. Hence, social behaviour and disadvantage is more than a question of individuals' motivations; it is also a reflection of the society of which they are a part.

Human nature is not biologically determined, but conditioned by social existence, and particularly by the need to cooperate within a technically defined division of labour.

Poverty now reflects inadequate socialization and/or the incompetent management of social, productive resources. Which suggests either improved conditions of life for the disadvantaged – welfare, health, education, housing, etc. – to improve people's identification with social ends and/or a competent, trained, technocratic elite able to better manage and plan society. However, both socialization and management are judged to be 'appropriate' and 'competent' with regard to the requisite social cooperation conducive with the extant technical division of labour; a division of labour that is constantly changing with technical progress.

Society is managed through social institutions that coordinate individuals' activity towards the achievement of social ends. And it is the role of development theorists to analyse the institutional structure for the management of society in the context of a changing technical basis for cooperation, and to suggest changes in the institutional management of society that might improve people's livelihoods and lifestyles. Development policy is oriented towards 'structural change' – *structuralism* (n.b. this is *not* 'structural adjustment').

Clearly, to be able to better manage resources in the social interest presupposes an understanding of how society functions – a model of society – and how individuals *could* be organized to take better advantage of society's resources and the future potentials of technical change.

To this end there is an extensive literature on 'measuring' development. By definition, judging development by the level of exchange between individuals, the value of transactions, as indicated by the gross national product, or national income per capita, while addressing the concerns of modernization theorists – individuals competing to realize their biological potentials – is inadequate for structural analyses that address the possibilities of social cooperation. The World Bank, which generally accepts a modernization theory approach to development (see Chapter 11) classifies 133 developing countries, from Albania to Zimbabwe by income per capita (World Bank 1996: 188–9). Perhaps the most fundamental criticism from structuralist theorists is that such an index ignores income variability within countries. Poor countries typically have an extremely rich small minority, and an educated middle class whose standard of living is considerably higher than the poor majority.

To assess development, taking account of such gross inequality, attempts have been made to construct some form of *standard of living* index. Summers and Heston (1990) calculated a purchasing power parity index in an attempt to provide a method to compare per capita income between countries. For the purposes of comparing, say, Rwanda and Italy, per capita GNP in Rwanda in 1994 was $80, and in Italy it was $19,300 (World Bank 1996: 188–9). What do these figures mean? Nobody could live for $80 for a year in Italy, so it is not a guide to comparative development and standards of living. Clearly the figures only have meaning in as far as there is some indication of what can be bought in Rwanda and Italy for $1, taking into account the different ways of life in these societies – purchasing power parity. Other elements that could be included in 'standards of living' development indexes include: inequality and distribution, infant mortality rates (deaths of children in the first year of life per thousand births), life expectancy, health service provision, access to potable water, education and literacy, households with electricity, and sanitation. On the issue of measuring comparative development, see Meier (1995), Chapter 1.

To try to produce a meaningful development index, the United Nations Development Fund tries to conceptualize poverty in a *human development perspective*, a combination of short life, illiteracy, social and economic exclusion, and lack of material means. A human poverty index (HPI) and a human development index (HDI) are calculated. The HPI concentrates on deprivation in the three elements that are considered to be essential to human life: longevity – the 'vulnerability to death at a relatively early age'; knowledge – exclusion from 'reading and communication'; and standard of living – overall 'economic provisioning'. The HDI includes: longevity – 'life expectancy at birth', educational attainment – 'literacy plus primary, secondary and tertiary enrolment ratios', standard of living – 'GNP per capita in terms of purchasing power parity' (UNDP 1997: 122, 125). The purpose is to identify the 'poorest' countries: those countries in most pressing need of assistance and structural change to achieve higher living standards..

While structuralist policy options, combining 'sustainable economic growth with social justice' (Stewart 1994: 48), implying a degree of equality or 'fairness', generally try to target those most in need – the poorest – the emphasis on poverty alleviation differs for each country depending on the institutional structure, and *who are* the 'poor'. Populist theories highlight particular disadvantaged groups – peasants, children, the unemployed, etc. – who have been adversely affected by processes of industrialization and 'modernization'.

Structuralist conceptions of sustainable development emphasize that the institutional structure should be compatible with the technical division of labour. In this regard some theorists stress that technical development must be appropriate to the social structure, emphasizing relatively small-scale economic activity. However, other interpretations assert that society has to change to meet the needs of technical change. But the *radical structuralists*, dependency theorists, are not so sanguine, being very sceptical that the institutional basis of development can ever be managed, as the political elite, representing vested economic interests, are able to stymie any progressive change towards equality.

These debates are addressed in Chapter 12.

Class struggle

The structuralist concern with the management of society, by 'experts', to achieve an improvement in the general standard of living, is based on the belief that there is a common interest between individuals. This assumption is the dynamic of the social democratic political project to achieve compromise and cooperation between individuals, through pluralist political institutions. In contradistinction, modernization theorists see individuals as incorrigibly competitive, with the objective of development policy being to create an environment of individual freedom and appropriate incentives, to maximize and realize individuals' potentials.

Individuals are either considered to be *dependent on* society or *independent of* society.

An emphasis on change as a consequence of class struggle combines the insights of structuralist and modernization development theorists.

A dialectical relationship is posited between individuals' 'modernization of the mind', and social experience and the exercise of power. Such an interaction is reflected in the concept of *conscientization*: people becoming aware of themselves through social life. People are creative beings, a creativity born of attempting to fulfil their potentials through social experience, by which people become aware of their abilities, talents and ambitions. And to be able to realize their potentials, individuals have to be conscious of the 'social, political, and economic contradictions' in society that frustrate people from fulfilling themselves, and they have to 'take action against the oppressive elements of reality' (Freire 1975: 15): oppressive elements that limit people's opportunities for living a full life.

People are not independent of, or dependent on, but interdependent *within* society: they differentiate themselves and become aware of their individuality through social interaction. And people become conscious of their potentials by identifying the social constraints and obstacles that frustrate their ambition. What is peculiar about humans as a species is that they are aware of *themselves*; humans are *self-conscious* animals (discussed in detail in Chapter 14). People can choose *today*, based on the experience of *yesterday*, how to better organize their activity *tomorrow*. And in as far as there is social change and development (progress), this will be a consequence of people's overcoming social constraints by being conscious of other people with shared interests, and organizing, or threatening to organize, or there is a portent of an organized movement, to challenge the relationships of power in society. These relationships underpin the social status quo, constraining disadvantaged individuals from fulfilling their potentials: the conflict is *political*.

Challenging the social status quo to mollify these constraints to individuals' creativity is fundamental to development (as opposed to change). Such a challenge to social power may be purposeful, such as the struggle in Nicaragua from 16 May 1926, when Augusto Sandino returned to organize the battle against US imperialism, to 17 July 1978, when the dictator

Anastasio Somoza ran for Miami; or the struggle, while being a reaction against oppression and the extant status quo, may be based on a conception of power and the practice of social change such that the transition to a new balance of power does not realize the promises and potentials of the new society, as in South Africa (Murray 1994); or the change may be so uncoordinated and spontaneous as not to have any specific objectives other than to remove the old order, becoming the precursor to a society where people's ability to fulfil their potentials is regressive, such as the changes in the Soviet Union and Eastern Europe 1989–91 (Andor and Summers 1998).

However, most organized change oriented towards development is not conceived in terms of a challenge to the social power structure. Indeed, development projects normally take existing relations of power as given, and are intended to allow the disadvantaged a little more opportunity to fulfil their potentials within the parameters of the status quo. However, 'Washing one's hands of the conflict between the powerful and powerless means to side with the powerful, not to be neutral' (Paulo Freire, see Freier 1975).

Such is the effect of the structural adjustment programmes sponsored by the World Bank discussed in Chapter 11.

Where progress is conceived of as people's greater opportunity and freedom to fulfil their social potentials, changes in society consequent upon such 'official' development policy only lead to development where inadvertently, and normally without the protagonists' being aware, the frustrations engendered by such policies and programmes create so much unrest that the authorities are forced to extend the parameters of individuals' freedom, and class power is surreptitiously challenged and redrawn.

John Walton and David Seddon detail how since the mid-1970s, internationally, a 'wave of price riots, strikes, and political demonstrations . . . swept across the developing world . . . reminiscent of [the] classical food riots . . . documented in European social history' (Walton and Seddon 1994: 23). And the catalyst that sparked these spontaneous social challenges to development orthodoxy – that is, development theory that seeks to ameliorate human suffering while essentially preserving the iniquitous status quo, or at least not questioning the source of advantage and privilege – is the increasingly integrated, globalized, national, economic relations of people's daily existence: a globalized world economy that has created such gross inequalities that 'rich' countries, with some 20 per cent of world population, account for 86 per cent of world consumption. And the richest 225 people in the world have a combined wealth of more than $1 trillion ($1,000,000,000,000), equal to the income of the 2.5 billion poorest people! Consequently more than a billion people do not meet the United Nations Development Programme levels for the basic consumption of food; around two and a half billion lack basic sanitation; one and a half billion have no safe drinking water; over a billion have no inadequate housing; etc. And it is getting worse (UNDP 1998). Even in the USA 'official' urban poverty increased from 14.2 per cent in 1970 to 21.5 per cent in 1993.

This trend towards social degradation, disadvantage and inequality, especially if placed in the context of the economic analysis at the end of Chapter 5, which heralds a world of declining profits and increasing exploitation to the advantage of an ever decreasing, international, capitalist ruling class, portends a future of conflict and suffering in the achievement of development.

> In the name of 'development', 'progress' or 'wealth-creation', 'economic adjustment', people all over the world are being subjected to constant upheaval and involuntary change. In the name of these noble abstractions, the landscapes of whole countries have been devastated, whole populations forcibly uprooted .
>
> (Seabrook 1993: 247)

Development theory, theorists, and consultants attempting to rationalize and apply Western conceptions of development, attempting to subordinate social change and environmental externalities to the existing, market-led accounting system, are part of 'the conjuring trick of the millennium . . . The unalterable laws of economics are pitted against the all too mutable fabric of the planet . . . [an enterprise imbued with a] profound intellectual dishonesty . . . [and which] is a mere ideological construct' (Seabrook 1993: 14–15).

There is a huge literature on conflict inherent in development: Jeremy Seabrook (1993) highlights, among others, case studies in India, Brazil, the Philippines, Indonesia, Great Britain and Thailand; Mamphela Ramphele (1991) addresses the problems of change in South Africa; David Reed (1992) includes case studies on Côte d'Ivoire, Mexico and Thailand; the Pacific Asia Resource Centre (1993) refers to case studies in Brazil, Central America, Jamaica, Sudan, India, Indonesia, the Philippines and Japan; Bruce Rich (1994) highlights the World Bank's activities in India and Brazil; Robin Broad (1988) discusses the policies of the International Monetary Fund and the World Bank in the Philippines; Johnson and Bernstein (1982) include studies on Brazil, Mexico, China, Mozambique, India, Ecuador, Chile, Colombia, Senegal, Jamaica, Ghana, Brazil, Morocco, Peru, Kenya and Bolivia; and there are also works by Walton and Seddon (1994), Pilger (1994, 1998), and Beckles and Shepherd (1996).

In recent times the issue of conflict and development has been considered under the question of the emergence of *civil society* (Burbridge, 1997; Tandon and Darcy de Oliveira, 1995).

The temptation is to despair that progress is a hopeless ideal, a paragon, the stuff of fantasies, and to turn to religion and evangelize to the world of the benefits of a utopia. Or to wash your hands of the convulsions of social life, dismissing people's ideals, ambitions and experiences as just so many discourses to be intellectually deconstructed (see Chapter 16).

The alternative is to take sides: to recognize that the dynamic of social life, social change, progress and development, is *power*.

[A]ny action that will achieve a positive change must be founded on an understanding of the need for a redistribution of power.

(Lappé and Collins 1977: 331)

The theorist, the academic, the student, or the consultant cannot be neutral. There is no fence to sit on.

The concept of power, in the context of understanding human potential to be a dialectic of people's biological inheritance and their social experience, centres on the concept of *class*, addressed in detail in Chapter 13. Class power reflects people's ability to control the social means of production, and hence social existence. The concept of class does not *describe* people's lifestyles, but *explains* their interests. And the class position of any one person might be ambiguous: for instance a wage labourer who may be a shareholder and therefore part owner of their place of work, and whose dividend income may increase with a rise in profits consequent upon his/her redundancy; or a cleric, who on the one hand propagates religious beliefs that are a 'fantastic reflection in men's minds of those external forces which control their daily life' (Engels 1975: 128), creating a passivity in the face of oppression. However, on the other hand, religious teachings might be taken literally (as with the *liberation theologists*), highlighting poverty in this world rather than heaven in the next. For instance: 'In a world where the poor were unimportant Jesus repeatedly gave the impression that they are all central to God's concern' (Vallely 1992: 144). See also Elliott (1989) and Alfred and Hennelly (1992).

Central to the analysis that the logic of class power underlies development and change is that to prosecute their class interest people have to be class-conscious. To act in accordance with this interest requires organization. The development theorist is at one and the same time a political activist. And the development strategy is one of *participation*. 'Collective leadership means leadership made by a group of persons and not by one alone or by some persons in the group' (Cabral 1979: 248): leadership oriented towards 'solving concrete problems in people's lives, making every effort to resolve them . . . [guaranteeing] that all citizens get effective support and non-bureaucratic solutions to their problems' (Machel 1975: 19). To do all this *and* to be participative is only possible if participation is conceived of as an emergent process, a process in which people themselves change. On the unfolding of such a process in Cuba, see Cole (1998), Chapter 3.

Class-consciousness and political organization are part and parcel of a process of self-realization: *praxis* (see Chapter 17).

Fundamental to development policy is the process of conscientization: 'none of us can live fully . . . as long as we are overwhelmed by a false view of the world and a false view of *human nature* . . . Learning how a system can cause hunger . . . becomes . . . a vehicle for a great awakening in our lives' (Lappé and Collins 1977: 329, emphasis added). And education is critical, where each student must become critical, understanding that the world is not 'given', but is a world being socially constructed, and thus a world that can be changed and transformed (Freire 1975, 1993; Hope and Trimmel 1995; Lappé and DuBois 1994; Wisner 1988).

On the theme of participative development see Friedman J (1979, 1992), Ghai and Vivian (1992), Fowler (1997), Brohman (1996), Farrington and Bebbington (1993), Hyden (1983), and Hippler (1995).

> The poor learn from . . . conflicts . . . and grow in consciousness and political power . . . In this process, the poor themselves define their own struggle . . . [and] development becomes radically participatory.
>
> (Wisner 1988: 15, 26)

Class-consciousness and development are addressed in detail in Chapter 13.

Development and human nature

Different perspectives on development interpret the same experience of social change differently, assessing such changes according to different conceptions of progress. The human facility to be self-conscious and choose is understood within distinct intellectual parameters.

Individuals' chosen activity can be understood to reflect their perceptions of *their* interests: after all *they* choose what *they* would like to do. And development is the extension of individuals' rights to behave according to their preferences. This is *modernization theory.*

All individuals achieve their ends through social experience, and their activity can be seen to reflect the social context of their behaviour. Individuals adapt to the need to socially cooperate. And development is a question of the better management of individuals' social responsibilities. This is *structuralist theory.*

People's idiosyncrasies can also be understood to reflect *and* affect their social experience. In such a dialectical relation the future is indeterminate, though *potentials* can be identified, the achievement of which is a consequence of people's awareness of the possibilities of change towards fulfilling their social potentials. This is theory of *class-consciousness.*

Respectively: individuals are *independent* actors, *dependent* participants, or *interdependent* partners.

11 Development as the fulfilment of individuals' potentials
Modernization

Modernization

> General backwardness, economic unresponsiveness, and lack of enterprise are well-nigh universal in the less developed world. Therefore if significant economic advance is to be achieved governments have an indispensable as well as a comprehensive role in carrying through the critical and large-scale changes necessary to *break down the formidable obstacles* to growth and to initiate and sustain the growth process.
>
> (Bauer 1991: 187, emphasis added)

What are these 'formidable obstacles'?

Just as for egocentric environmentalists and subjective preference theory economists, *modernization theory* development theorists see progress as a consequence of the free choice of individuals to act according to their own, best interests. Development *is* the extension of individual freedom: individuals have the right to decide what is best for themselves. Social life is guided by individual responsibility: and development policy is about promoting changes in individuals' aptitudes and attitudes so that they become more attuned to recognizing the potentials in alternative courses of action for improving their own lives.

But not only do individuals have to modernize their minds, societies also have to modernize, to allow people to freely choose. The utopia envisaged is a society where individuals are

> free standing agents, each equipped with his or her distinctive set of preferences, and each transacting on an equal footing with whosoever it pleases him or her to transact with.
>
> (Thompson 1992: 182)

On such a conception of utopia, see the section 'The neoclassical utopia' in Chapter 3.

Social institutions that are considered to be adverse to individuals' material progress, those institutions that inhibit the achievement of a perfectly competitive utopia, have to be reformed. This includes: institutions such as state agricultural marketing boards, which set the prices for cash crops, thereby frustrating the competitive dynamic of agricultural markets; tariff protection for national industries in the name of employment creation and protection, which insulates enterprises from international competition, and thereby limits the incentive for firms to improve their productivity, inevitably leading to larger and larger economic subsidies prejudicing individuals' free choice; examples of political favouritism, such as crony capitalism in East Asia highlighted in Chapter 9, which again limits free individual choice to the advantage of powerful people; and the activities of trades unions, which in a similar way distort markets in favour of particular vested interests.

Between the sixteenth and eighteenth centuries in Europe, society went through a phase called the Enlightenment, when people's behaviour became more and more based upon rational calculation than superstition, and fatalistic attitudes towards the use of the natural environment and the improvement of living standards were replaced by control through scientific investigation, leading to purposeful change.

> The world view and value system that lie at the basis of our culture ... were formulated ... [b]etween 1500 and 1700 [when] there was a dramatic shift in the way people pictured the world and in their whole way of thinking. The new mentality and the new perception of the cosmos gave our Western civilization the features that are characteristic of the modern era.
>
> (Capra 1982: 37)

In sum, individuals acquired the attitudes and intuitions to offset the costs and benefits of activity, to maximize the returns to themselves of economic behaviour. As people became aware of themselves as independent beings, endowed with tastes and talents, development 'took off'.

> The first set of assumptions shared by modernization researchers are certain concepts drawn from European evolutionary theory ... social change is unidirectional, progressive, and gradual, *irreversibly moving societies from a primitive to an advanced stage*, and making societies more like one another.
>
> (So 1990: 33, emphasis added)

Change is 'unidirectional' because all individuals are (believed to be) independent beings, and the inevitable self-realization of individuals' interests will precipitate a movement towards a 'free' society: 'the slave's continuing desire for recognition ... was the motor which propelled history forward' (Fukuyama 1992: 198). People strive to be themselves. And such a change is progressive, as it is in accord with individuals' nature, and gradual, because it is never in the interests of any one individual to live in a 'free' society –

that is, a society free for everyone else – since people are always alert to their personal preferences, tastes and advantage. Such a freedom for all has to be politically protected, but is always precarious, as people in positions of power are instinctively tempted to further their own interests. After all, their own experience is the only reality they actually know.

From the traditional to the modern

In development terms, progress is a change from a traditional to a modern society:

* from a society characterized by little economic specialization, a limited division of labour, marginal market activity and essentially subsistence production and where the capitalistic organization of production is unusual, with a very small wage-labour force; from a culture steeped in tradition, hierarchy, and obstacles to change; from a high birth rate reflecting poor health facilities and the need for support in old age; from a rural–urban flow of goods; to
* a society with a highly specialized division of labour, where market exchange is widespread and people on the whole do not produce what they consume; where entrepreneurs are pivotal in organizing production, and most people earn wages; where there is a low birth rate as individuals adopt attitudes of individual responsibility consequent upon market competition, and as a basic welfare safety net is provided by the state; where there is a rural–urban exchange of goods; where social institutions are continually adaptive to individuals' changing needs, but also forcefully protect individuals' freedoms from vested interests; 'there is an unquestionable relationship between economic development and liberal democracy' (Fukuyama 1992: 125).

The social context of economic life is pivotal in this process of modernization.

> Cultural obstacles for change are common in all societies, for the obvious reason that an impending transformation threatens existing habits, ways of life and social prejudices . . . To respond to change might mean altering one's own priorities, educational system, patterns of consumption and saving, *even basic beliefs about the relationship between the individual and society.*
>
> (Kennedy 1993: 17, emphasis added)

As a perspective on development this goes right back to Adam Smith in the eighteenth century. But more recently, modernization theory, as a basis for development policy intended to create societies that maximize the freedom of individuals, has its roots in the work of Walt Whitman Rostow. In *The Stages of Economic Growth: An anti-communist manifesto* (Rostow 1960) Rostow outlines a belief that *all* societies have evolved or eventually will evolve, through five stages: traditional; preconditions for 'take-off'; the take-off; the drive to maturity; finally reaching the apogee of human development, the 'end of history' (Fukuyama 1992: Part V) – the age of high mass consumption.

'Traditional society' is characterized by a limit to the standard of living reflecting the fact that: 'the potentialities which flow from modern science and technology . . . [are] either not available or not regularly and systematically applied' (Rostow 1960: 4).

The 'preconditions for take-off' are an age in which scientific techniques and insights begin to be applied systematically to production processes: the potentialities of modern science increasingly become available. The improvement in efficiency in production necessarily expands the technical division of labour, and the scope of market exchange. And enterprise emerges. People who are willing to take risks in the pursuit of profit increasingly dominate economic affairs, but their entrepreneurial intuition is constrained by 'traditional low-productivity methods, [and] by old social structures and values' (Rostow 1960: 17).

Then, eventually, all these preconditions come together and there is take-off.

> The beginning of take-off can usually be traced to a particular sharp stimulus. The stimulus may take the form of a political revolution which affects directly the balance of social power and effective values, the character of economic institutions, the distribution of income, the patterns of investment outlays . . .
>
> (Rostow 1960: 36)

But the effect is that the competitive dynamic of market relations of exchange ensures that the latest scientific knowledge is applied to production, as entrepreneurs compete to be profitable. Levels of investment increase, and certain leading manufacturing sectors emerge, which are characterized by high growth rates, and a political, social, and institutional framework evolves that serves to consolidate market exchange and private enterprise.

> It is evident that the take-off requires the existence and the successful activity of some group in society which is prepared to accept innovations.
>
> (Rostow 1960: 50)

Enterprising individuals must have an incentive to apply the latest technology to production: and that incentive is profit. And these dynamic individuals must have a direct interest in organizing production and supply, in order to respond to consumers' demand: the entrepreneur is pivotal to development. So the policy of modernization is essentially the promotion of private enterprise and capitalism. And progress, economic development, is a consequence of the benefits of modernization trickling down from the enterprising rich to the indolent poor, as the latter are productively employed by the former.

Remember, it is not 'from the benevolence of the butcher, the brewer, or the baker that we expect our dinner, but from regard to their self-interest' (Adam Smith, quoted Heilbroner 1992: 53).

The 'Third World country ... represents, in effect, an earlier age of mankind' (Fukuyama 1992: 130). Referring to the 'uncivilized nature' of the Cherokee tribe of Indians in North America, Senator Harry Dawes of Massachusetts drew attention to the communal nature of their way of life: 'They have got as far as they can go, because they hold their land in common ... There is no selfishness, which is at the bottom of civilization' (Jones 1996: 96).

As a strategy for promoting the modernization of backward, poor societies, in the 1960s *dualism* was the favoured option. The popular image of developing societies was one of a very small, advanced, modern, industrial sector, juxtaposed beside a large, backward, traditional, agricultural sector. And in 1954 Arthur Lewis posed the problem of development as: how to make the traditional, backward sector more like the modern, advanced sector; and, more particularly, how to inject market rationality into the behaviour of traditional, agricultural producers. The answer was *import substitution industrialization* (ISI). Commodities that had hitherto been imported would be produced locally, and to be able to compete with foreign imports, local producers would be protected by tariff barriers, which would make imports prohibitively expensive. As local industrial production expanded, individuals would find that they could earn more in a factory than in agriculture, and as labour shortages appeared in the agricultural sector, farmers would have to improve efficiency to be able to offer higher wages to retain labour. And the traditional sector would be modernized. A naïve belief underlay the theory of dualism that market prices respond to consumers' demands, so that producers are paid what they are worth: a belief called into question by the theory of the second best and the capital controversy in Chapters 3 and 4. But in the 1960s and 1970s policies of ISI did seem to be having some success, particularly in Brazil – the so-called 'Brazilian Miracle'.

> [I]mport-substituting industrialization was the only possible strategy, because initially the domestic market was the only market which could be secured by state intervention, for domestic industrial production.
>
> (Sender and Smith 1988: 72)

However, increasingly, declining imports of consumer goods were replaced by the import of producer goods (raw materials and technology, etc.), leading to balance of payments problems, and a slowing of economic expansion. And ISI was associated with 'state interventions to secure short- and long-run availability of adequate flows of foreign exchange to finance the inevitable increases in import requirements' (Sender and Smith 1988: 67). State intervention and economic management was a *sine qua non* for development.

Structural adjustment

In the 1980s and 1990s the policy agenda to create a free society as a path to development has increasingly been interpreted as the creation of free *markets,* not subject to the economic management characteristic of ISI development strategies: free markets to which individuals will spontaneously adapt in the desire to maximize personal utility, so that their enterprise will beget social progress.

> Those of us who have been fortunate enough to be born in the United State in the 20th century naturally take freedom for granted ... But freedom is very far from being the natural state of mankind: on the contrary it is an extraordinarily unusual situation.
>
> (Friedman 1976: 8)

And the key to this elusive freedom? It is individuals having control over their own *economic* destiny.

Everyone serves their own interest: everyone is an independent individual, and at least in free markets people can buy and sell according to their own preferences. As soon as politicians begin to allocate and manage resources in the general interest, inevitably, in one way or another, individuals are coerced to abide by the politically defined 'general will'. Even taxation defines how people *should* spend their income and order their lives. Certainly governments and political authorities have a pivotal role in ensuring economic development: fostering market institutions such as an efficient banking system, protecting market institutions by guaranteeing law and order and enforcing private contracts, maintaining national sovereignty through defence expenditure, etc. The state must act as an umpire in market relations of exchange between independent individuals, but the more the state assumes the role of economic guardian – allocating resources in spite of individuals being able to look after their own interests and freely choose in the marketplace – then this is the road to conflict and coercion.

> [E]conomic freedom is an end in itself ... competitive capitalism ... *promotes* political freedom because it separates economic power from political power and in this way enables the one to offset the other.
>
> (Friedman 1962, quoted Heilbroner and Ford 1971: 228–9, emphasis added)

In contemporary times the mantle of modernization theory, the creation of a free society in which entrepreneurs can prosper, has passed to the International Monetary Fund and the World Bank in the guise of *structural adjustment programmes* (SAPs). From the policy emphasis of the 1960s and 1970s, that efficient expert intervention under the auspices of the political authorities was an essential component for the rational management of resources for effective economic development, it became anathema to economic and development theorists that the state, through ISI or any other interventionist policy, could manage development.

> [T]ake Ghana ... as a case in point ... Ghana's losses on public sector enterprises were so large as to contribute significantly to budget deficits; monetary expansion to repay the domestic deficit was contributing to rapid inflation; overvalued exchange rates, negative real interest rates and price controls were giving rise to extensive rent-seeking [speculative] behaviour; public employees were 'moonlighting', neglecting the duties of their employment and professional people were streaming abroad – there was hardly any area influenced by *public policy* that was not in distress.
>
> (Cassen 1994: 7, emphasis added)

The change in economists' vision from economic management to free markets has already been addressed in Chapters 3 and 4, but part and parcel of this sea change in economists' and development theorists' intuitions has been the burgeoning debt crisis. In 1982 the Mexican balance of payments crisis meant that there was not enough foreign currency to meet all the claims from foreigners for payment for imports, loans, and international services. And Mexico defaulted on repayments on international debt. To acquire foreign exchange for international transactions the Mexican government had to turn to the International Monetary Fund (IMF), but as a condition of the loan Mexico had to change economic priorities and institute policies to stabilize the economy, with the aim of achieving a surplus of foreign exchange to repay the loan. These conditions are known as an SAP (structural adjustment programme), and are essentially geared towards the creation of a viable, free-market, capitalist economy.

SAPs are the application of neoclassical economic theory to development. It is a question of relative prices being adjusted to alter the allocation of resources to move towards economic equilibrium, and the maximization of consumers' utility.

> A structural adjustment program can be defined as a set of measures that attempts to permanently change relative prices of tradable to non-tradable goods in the economy, in order to reallocate, or help reallocation of the productive factors in accordance with a new set of external and internal economic conditions ...
>
> (Edwards and Van Wijnbergen 1989: 1482)

'Non-tradable goods' refers to domestic production, and SAPs are intended to relocate resources into the tradable goods sector of the economy, so as to increase exports. And there are other policies intended to depress demand and reduce imports, and thereby realize a surplus on the balance of payments with which to repay the loan of foreign exchange from the IMF. 'The Fund without doubt favours market related solutions as the more efficient means of correcting deficits' (Killick 1984: 179).

But there has not been the predicted renaissance of economic activity in the poor, indebted economies, unambiguously leading to economic development and social progress. Of course, as we saw in Part II, different economic perspectives define economic reality differently, placing distinct intellectual parameters on economic, and hence development, issues.

Within the subjective preference theory of value and the modernization theory of development the assumptions that underlie structural adjustment policies are presented as objective, and the approach scientific, and the language used by development economists, which presents the issue as a technical problem of merely adjusting relative prices (see Edwards and Van Wijnbergen above), enhances the aura of objective correctness. 'Their own market interventions, those promoting exports, became 'incentives'; interventions leading in any other direction were called 'distortions' (Broad 1988: 231).

> Evaluating the effects of ... structural adjustment programmes ... depends on the method selected ... [A]re the disappointing results to be attributed to the poor execution of the programme or to the programme itself? ... [And] it is not clear how much time has to go by before the adjustments affect macroeconomic development ... [and] the indicators ... used to evaluate the effects ... are disputable.
>
> (Lensink 1996: 97)

But the 'disappointing results' were not confined to the economy.

> [T]he link between neo-liberalism [structural adjustment policies] and Latin America's increased poverty and inequality can be all too obvious. In Bolivia the redundancy notices issued to thousands of factory workers since the government began its structural adjustment ... in other cases ... the connection is less direct ... [the fall in] international commodity prices, the end of commercial banking in the region [etc.] ... [The NGO] Christian Aid believes that 'Throughout the third world ... Structural Adjustment Programmes spell hardship for people in every aspect of their lives – health, education, work, culture'.
>
> (Green 1995: 94)

The economic policy of 'permanently changing the relative prices of tradable to non-tradable goods' meant that the priority in economic activity was to achieve a surplus on the balance of payments, and the end of development programmes specifically targeted at the old, young and sick: the priority was enterprise. Workers were obliged to exercise wage restraint in order to reduce costs, 'in countries where the difference between wage and peonage is slight' (Pilger 1994: 195). In the Philippines, which was 'structurally adjusted' in the early 1990s, hundreds of thousands of workers lost their jobs as industry was rationalized. And tens of thousands of children died silently and unnecessarily as government spending was slashed, and food subsidies and milk and vitamin supplements for children were withdrawn (see Pilger 1994, Chapter V).

> [S]tructural adjustment policies ... typically create an economy subor-
> dinated to the needs of foreign investors and a two-tiered society, with
> islands of great privilege in a sea of misery, sometimes called 'economic
> miracles' if investors benefit sufficiently.
>
> (Chomsky 1996: 82)

For modernization theorists, economic growth – a consequence of individual freedom – *is* development.

> [For the World Bank] economic growth ... remains the unquestioned
> *imperative of development strategy* and that the benefits derived from
> economic growth can best provide the means for addressing ancillary
> concerns [such as poverty relief, environmental protection, etc.].
>
> (Reed D. 1992: 4, emphasis added)

Although 'experts' might dispute theoretical assumptions, relevant episte-mologies, and appropriate analytical priorities in the achievement and measurement of development, to the unemployed, the uneducated, the sick, and the dying, measures of development are not quite so disputable. Modernization theorists are not oblivious to this mounting suffering created in the name of development, and have been concerned to explain why these programmes have not delivered something that could unambiguously be defined as progress, even within their own intellectual parameters. The problem is being addressed by a school of thought known as *new institutional economics* (Harriss *et al.* 1995; North 1990; Stiglitz 1992).

It is thought the problem lies in that individuals are not able to act according to their nature: they are prevented from 'being themselves', from being enterprising and dynamic. It is argued that people's behaviour is condi-tioned by their understanding of how the world functions. But

> information is incomplete [for the achievement of 'equilibrium'] and
> there is limited mental capacity by which to process information ...
> Individuals possess mental models to interpret the world around them.
> These are in part culturally derived ... In part they are acquired through
> experience ... and as a result there are different perceptions of the
> world ...
>
> (North 1995: 18, 19)

The lack of coincidence of perception creates uncertainty, so that people – to try to create a degree of predictability, to try to inject some order into their lives – interact through institutions. However, these institutions become a critical constraint on individuals' freedom, because 'individuals ... with bargaining power as a result of the institutional framework have a crucial stake in perpetuating the system' (North 1995: 20). Change according to individuals' preferences is frustrated.

The trend towards the modernization of society and the liberalization of the economy is inevitable, albeit conditioned by particular institutional constraints in each society. For instance, in the analysis of development in Cuba, because for decades individuals' subjective preferences have been denied by a planned economy and fixed prices – institutionally created market imperfections – market prices and economic transactions have not been able to adjust to reflect individuals' opportunity costs, and hence their preferences for utility. For instance, equilibrium contracts in the factor (productive input) market have been frustrated. Wages have not reflected

> the high opportunity cost of time [the alternative way in which individual producers could utilize their time]; the increase in the number of households due to the decreased role of extended family structures [consequent on the increased role of the state in ensuring social welfare]; the increase in the households with multiple earners. Of these factors the last two are relevant in the Cuban economy.
>
> (Bettancourt 1991: 19)

This is an application of new institutional economics to a (potential) Cuban process of modernization (Cole 1998: Chapter 5): an analysis in which the issue of returning assets nationalized after the revolutionary victory of 1 January 1959 to their previous owners would also be addressed (Thomas 1992; Mueller 1991).

The analysis is based on a set of *beliefs*, as articulated by Samuel Huntington, Professor of the Science of Government and Director of the Olin Institute of Strategic Studies at Harvard University. The principles of liberty, democracy, equality, private property, and markets, a ' set of universal political and economic values' that constitute the 'national identity' of the United States, requires that America occupies a position of 'international primacy' for the collective benefit of the world.

> Since this is a matter of *definition* [a *belief*] ... evidence is ... irrelevant. In evaluating Washington's promotion of human rights, for example, we may put aside the close correlation between US aid (including military aid) and torture demonstrated in several studies, running right through the Carter years ... Such considerations are the province of small minds, unable to appreciate Higher Truths.
>
> (Chomsky 1996: 27–8, emphasis in original)

Modernization and good governance

So, Rostow's inevitable evolution towards modernity, the state in which we can all benefit from the enterprise of gifted, calculating, self-interested individuals, is thwarted by institutional constraints that maintain the status quo to the advantage of vested interests. Hence development policy should now be orientated towards overcoming the cultural and institutional constraints that militate against individuals' freedom to choose, constraints that are particular to each social context.

To this end, in the early 1990s the World Bank emphasized *good governance*, defined as government that establishes

> the rules that make markets work efficiently and . . . correct for market failure . . .
>
> (World Bank 1992: 1)

On the issue of good governance see Goetz and O'Brien (1995), Hewitt de Aleantara (1998), and World Bank (1992).

The problems and contradictions of defining economic policy in terms of market failure were addressed in Chapter 4 and will not be repeated here. Since the mid-1970s the policy priorities of the IMF and the World Bank have become closer and closer. The IMF and the World Bank were instituted at the July 1944 conference at Bretton Woods, New Hampshire (USA), when Western governments agreed a set of rules and institutions to regulate the post-war global economy, control conflict, and encourage economic growth. The IMF was to institute a regime of financial management to oversee monetary exchanges between member countries, and to guarantee sufficient credit for financing international trade and investment. The World Bank was to address the difficulties of developing countries participating in a competitive world market: 'the Bretton Woods system was . . . crucial to the overall stability of the whole post-war *liberal* system . . .' (Gadzey 1994: 143, emphasis added).

The Bretton Woods conference was about power in the world economy. The only national economy that benefited from the Second World War was the American economy. And the United States government was anxious to institutionalize US economic power, and 'the call for free trade was a diplomatic invitation to the British to give up their discriminatory system of imperial preferences [based on the British empire] . . . [which were] the main impediment to the expansion of American trade' (Gadzey 1994: 100).

As the number of countries achieving political independence in the post-war period grew, the loan portfolio of the IMF became more orientated to the developing economies, characterized by 'slow growth and an inherently weak balance of payments position which prevents pursuit of an active development policy [and hence economic growth]' (IMF 1974), requiring relatively long-term loans. In 1986, in the wake of the debt crisis, a specific Structural

Adjustment Facility for developing economies was introduced, followed by the Extended Structural Adjustment Facility in 1987, targeted at particularly indebted economies; both were intended to address the transition of developing economies into full market economies.

And the World Bank came to recognize that individual projects contributed little to development, conceived as expanded market exchange with the extension of individual consumer choice, in a setting of poor (imperfect) macroeconomic policies. So with the emerging debt crisis in the 1980s, attributed to excessive state intervention and market imperfections, reforming, structurally adjusting, the economic environment assumed priority.

In the 1980s and early 1990s the policy emphases of the IMF and the World Bank were distinct.

> As one Bank official admitted, 'In 1980 the [debt] problem was seen as a medium term issue, one which many thought would have been resolved by the mid-1980s'. . . Bank staff soon concluded that even BoP [Balance of Payments] improvements . . . would depend largely on domestic policy and institutional reforms . . . The Fund would continue to give short-term loan support . . . World Bank SALs [Structural Adjustment Loans] . . . would focus on restructuring economies through institutional and policy reforms . . .
>
> (Reed D. 1992: 13)

But as the debt crisis has deepened, the policy priorities of the IMF and the World Bank have merged: 'With both institutions active in the structural adjustment business, the old demarcation lines – short term and macro for the Fund, long-term and micro . . . for the Bank have tended to fade' (Polak 1996: 17). And 'a merger makes sense, and in time it will happen' (Crook 1991: 48): a trend that is reflected in the policy of the World Bank towards the transition economies of the former Eastern bloc, where the goal is 'the same as that of *economic reforms elsewhere*: to build a thriving market economy capable of delivering long-term growth in living standards' (World Bank 1996: 1, emphasis added). For a critical analysis of the activities and policies of the World Bank see Caufield (1997) and Gibbon (1992).

The Multilateral Agreement on Investment

The aim is to liberalize the world economy, in order to further the modernization, and hence the development, of those societies engaged in international trade. The General Agreement on Trade and Tariffs (GATT), the forerunner of the World Trade Organization (WTO), another Bretton Woods Institution for managing the world economy (although instituted in 1947, two years after the IMF and the World Bank), has been engaged in a process of deregulating the world economy to augment the competitive dynamic of international exchange. The last round of negotiations, the Uruguay Round, so called because it was launched in the South American resort of Punta del Este in Uruguay, took eight years to reach fruition (1986–94), after which GATT metamorphosed into the WTO, extending the area of responsibility to include services and intellectual property. According to Oxfam, 'the Uruguay round brought worsening conditions and unilateral and indiscriminate liberalization destroying poor people's livelihoods and tore at the fabric of social cohesion' (Oxfam 1998: 4).

Included in the Round was a proposal for investment liberalization. This was aimed at reducing restrictions on foreign investment, such as Taiwan's controls on 'highly polluting industries', or restrictions on insurance and banking investment in the Philippines and Thailand, or limitations on the purchase of land around nature reserves or national borders in Brunei, Pakistan and Brazil, or any national legislation intended, say, to protect child labour from exploitation, or environmental despoliation. The logic was to increase competitive investment in a broader range of countries, with the expectation of more employment, augmented economic growth, and higher GDP per capita: a Pareto improvement in the world economy (the contradictions of which have already been addressed in Chapters 3 and 4).

Such an investment liberalization strategy was strongly opposed by the developing countries, and discussion was shifted to the Organization for Economic Cooperation and Development (OECD), the Paris-based club of 29 rich countries, and the International Chamber of Commerce, where negotiations could be carried on behind closed doors without the troublesome intervention of the developing countries, who tended to put national interests before the concerns of transnational corporations. Known as the Multilateral Agreement on Investment (MAI), discussion has focused on three main issues:

1   Foreign investors must be treated as well as, or better than, domestic companies. This means that governments would not be allowed to protect 'infant' industries. Local industry '. . . typically disadvantaged by a relatively weak domestic infrastructure . . . [and] unable to produce on a massive scale and spread their costs across global markets' (WDM 1997a: 3) would have to compete with transnational corporations.

2   There can be no 'no-entry' restrictions. About 75 per cent of foreign investment takes the form of acquiring domestic companies. Almost invariably the subsequent reorganization and rationalization leads to job losses and reduced exports, as managerial decision making reflects the corporate strategy of the transnational corporation. To try to develop an industrial strategy for development, some countries, such as Kenya, Indonesia and Malaysia, limit the extent of the foreign control of national enterprises. Under the MAI, foreign investment would be allowed in any sector (except defence), and governments would be unable to favour domestic industries over international investors.

3　There can be no conditions on foreign investors. No requirements to meet: local employment targets; limit currency speculation; a minimum period for investment; minimum local content rules; requirements for the transfer of technology; etc. Hence there would be: no sovereignty of national laws; no protection of labour or of consumers; no environmental protection, 'which would legalize . . . a whole host of terrible and inhumane environmental practices, such as the use of drift nets or the clear-cutting of old growth forests' (Kellogg and Whitney 1998: 10); no control of restrictive business practices; etc.

In sum, no national development policy.

The aim is to extend the ability of transnational corporations to move finance rapidly around the world to maximize profits, regardless of the costs to workers, the poor, or the environment. The overall purpose is to restrict governments and the state in developing countries in order to exploit the business opportunities of an increasingly globalized world economy, and to heighten competition as the trend towards economic crisis deepens, ensuring that the least efficient enterprises absorb the losses, protecting the future profitability of the transnational corporations (see Chapter 5). Already the combined income of the top eight transnational corporations exceeds that of half of the world's population: 2.4 billion people (Kimber 1998).

Developing countries have not been included in the negotiations, and yet, if they accede to the MAI, they will be subject to binding decisions requiring them to change laws, or to be taken to an international tribunal and sued by transnational corporations for past or potential future damages, which could run into millions of dollars. Potentially any regulations that delay or prohibit foreign investment can be judged to be an expropriation of future profits to be compensated by governments. And once governments sign up to the MAI, withdrawal is prohibited for five years, and countries are bound by the agreement for fifteen years.

> *The MAI negotiators are pursuing a 'liberalization at all costs' approach without adequate analysis of the social, developmental and environmental implications.*
>
> (WDM 1997b, emphasis in original)

But why should developing countries choose to sign up to the MAI, and bias their development policy towards transnational capital? Why should they risk being locked into a subservient role of producing commodities and supplying cheap labour for foreign companies?

The MAI is being heralded as a stamp of approval; poor countries are being told that they will not receive foreign investment until they sign up.

> While foreign investment increased five fold in the decade to 1995, little of it goes to the poorest countries. Of the $112 billion investment in

developing countries in 1995, over 80% ended up in 12 countries. The 48 least developed countries with 10% of the world's population attracted only 0.5% of global investment.

(WDM 199a: 4)

The MAI was due to have been signed at the OECD ministerial meeting of 27/28 April 1998. However, an international campaign, including over 50 NGOs from twenty-five countries, trades unions, elected members of parliament, including the European Parliament, as well as government objections from the United States, France, Canada and the Czech Republic (although some of these objections were that the MAI is not liberal enough!), led to the ratification being postponed until at least January 1999, while the agreement is redrafted. However,

> despite growing opposition, negotiations will continue. Ministers announced a period of assessment and further consultations ... There have been recent proposals to change the IMF's Articles of Association [to include the MAI], push the MAI into the World Trade Organization, promote signing the MAI in renegotiation of the Lomé Convention, and include the MAI's provisions as a condition for developing countries to benefit from aid under the US African Growth and Opportunity Act.
>
> (WDM 1998)

However, the trend towards greater capital controls and regulations on foreign investment in Indonesia, Malaysia, Hong Kong, Brazil, and Russia, to stave off capital flight, currency devaluations and volatile investment flows, consequent on the deepening world economic crisis in autumn 1998 (see Chapter 5), may make the MAI less of a potential, as clearly many governments would resist the pressure to sign up to the MAI and allow foreign capital to flow in and out of national economies without restrictions. This is a reality recognized by British prime minister Tony Blair in his speech to the New York Stock Exchange (21 September 1998), when he called for a 'fast track new world financial order', with the IMF and the World Bank being reformed to pursue high rates of growth and development. This 'order' would not be left to the dictats of the rich countries, but would also be controlled by the developing economies and outside experts, and would be able to sanction – in consultation with governments – capital controls to combat the effects of currency speculation, offsetting volatile currency flows.

12 Development as fulfilling the technical potentials of cooperation

Structuralism

Structuralism

> The primary objective of long-run development can be summarised in very general terms as being sustainable economic growth combined with social justice ... Development involves *structural transformation* of the economy and society. Starting from low-productivity, largely subsistence agrarian structures, in the course of transition towards modern economic growth, the industrial sector grows in significance with the agricultural sector playing an essentially supportive role ... developing countries acquire *technological* capability and move away from primary specialization. During this process of transformation special efforts are needed to ensure ... *social justice* ... [with] a wide spread of economic opportunities and an appropriate *distribution* of social goods.
>
> (Stewart 1994: 48, emphasis added)

Development is not now merely a consequence of changing individuals' attitudes, of people developing an awareness of their own interests, with the benefits trickling down in a modern society from the privileged to the disadvantaged. Development is now a process of technical transformation. Certainly, production is based upon an improving technological capability, as society moves away from primary product specialization and industrializes, but in a context of fairness, and an appropriate distribution of social goods. Development is not a consequence of free enterprise and is more than modernization.

Social change, and the change in individuals' behaviour, is reinterpreted, without the stress on the extraordinary talents of enterprising individuals – entrepreneurs. The changes are structural (not 'structural adjustment'); changes in society to which individuals *have* to adapt. Society is more than the sum of the individuals of which it is composed. There is an evolutionary development dynamic beyond individuals' preferences, and the structural transformation of society is characterized by both an improved technological capability *and* social justice, both of which have to be managed by 'experts': economists and development theorists.

As we saw in Chapter 4 individuals can be conceived of as being technically dependent, and the technical division of labour sets the parameters to
individuals' behaviour and to social life. Individuals are not free to choose,
and subjective preferences cannot be the economic and development dynamic.
Development is a consequence of the efficient management of resources in
the general interest: that is, resource and social management relevant to the
prevailing technical division of labour. And to encourage and facilitate
economic cooperation between dependent individuals, society has to be organized through appropriate social institutions; the concern is not to facilitate
competition between independent individuals.

> Policy recommendations centre on finding ways in which *governments*
> can intervene to help private producers change . . .
> (Hunt 1989: 122, emphasis added)

Necessarily, to survive, everyone in some way relates to the productive
economy, and individuals are part of technically and distributionally defined
interest groups: miners, farmers, teachers, students, householders, the police,
etc. And the institutional, social framework is the mechanism through which
people from different interest groups arrive at compromise in the general
interest: pluralist, social democracy.

Development, then, is 'a multidimensional process involving major
changes in social structures' (Todaro 1981: 70), and development theory
has to address the particular, institutional constraints to changing technical
and economic cooperation in each society. That is: institutions have to be
appropriate to the technical needs of producers and the social imperative of
fairness. Just as for the cost-of-production theory of value, structuralist
development theories have to be sensitive to *actual* forms of social
behaviour. Analysis has to be empirically accurate and realistic rather than
based upon the utopian ideal of a modern society. Often, theorists within
the structuralist perspective on development make great play of the fact
that they deal with the real world rather than trying to achieve an unrealistic ideal. For instance, to answer his critics André Gunder Frank asserts:
'*Real world* system evolution has never been guided by or responsive to
any global . . . thinking or policy' (Frank 1996: 45, emphasis added), as
if the 'real world' is obvious to those who take the trouble to look (see
Part IV).

There is now not one fundamental obstacle to people being themselves
and frustrating development – a lack of individual freedom and consequently
free exchange – but potentially a plethora of institutional obstacles to effective cooperation and social harmony.

> The one clear conclusion is that there is no single cause of poverty
> among today's developing nations.
> (Gillis *et al.* 1983: 37)

In Chapter 4 we saw that the institutions of the economic system are intended to engender cooperation between producers, and that competition is important in as far as enterprises have to adopt the most efficient means of producing commodities, and therefore lower production costs, tending towards equilibrium prices. But competition, by which individuals attempt to maximize their own advantage, is acceptable only if individuals' particular interests do not contradict the broader general interest of full employment and a degree of fairness: technical efficiency and distributional equality. Remember, economic cooperation is unlikely to transpire where there are extremes of wealth and poverty. And anyway, with the economic and development dynamic no longer being independent individuals, there is no moral exigency towards the competitive market socially valuing individuals' talents and hard work – concerns echoed by structuralist development economists.

> The structuralist approach attempts to identify specific rigidities, lags and other characteristics of the structure of developing economies that affect economic adjustments and the choice of development policy ... A common theme in most of this work is the *failure of the equilibrating mechanism of the price system* [market exchange] to produce steady *growth* or a desirable *distribution* of income.
>
> (Chenery 1975: 310, emphasis added)

For instance, in the case of addressing the implications of globalization, the United Nations Conference on Trade and Development (UNCTAD) recommends a 'new' management strategy reflecting the changed technical relations of production that have given rise to transnational corporations, improved world communications, and the globalized world market.

> The most effective way of reducing poverty and increasing welfare in developing countries, is to *design* economic policies *explicitly around these objectives*.
>
> (UNCTAD 1996: 110, emphasis added)

The emphasis is not on free markets.

Distribution and populism

Once again, the questions of technology and growth, and of fairness and distribution, which were addressed in Chapter 4 under the cost-of-production theory of value, reappear: development priorities that may be frustrated by the 'equilibrating mechanism of the price system'. Structuralist analyses of development almost invariably focus on specific problems identified with elements of unfairness or inequality: questions of *distribution*, and the failure of industrial development to mitigate the problems of the poor.

And in as far as capitalist industrialization goes hand in hand with the decline of small-scale manufacturing industry and agriculture, populist (and neo-populist) ideologies rationalize the plight of unemployed workers and bankrupt individual entrepreneurs, and peasants. However,

> efforts to develop such ideologies of opposition to capitalist industrialization were made not by peasants or workers themselves but by urban intellectuals . . .
>
> (Kitching 1982: 20)

Experts, not individuals instinctively responding to relative prices, market signals, are the dynamic of change.

Inevitably a 'pre-capitalist' *utopia* is posited: 'a world of individual small-scale enterprise located in small towns and villages' (Kitching 1982: 22), where competition is mediated by community and family obligations. A relatively egalitarian distribution of income generates sufficient local demand to ensure employment. Consequently, populist ideologies are typically against extremes of wealth and poverty. Opulence and luxury consumption give employment only to a 'labour aristocracy'.

Populism has its roots in the industrial revolution, and the call by the Ricardian socialists and the Owenite cooperative movement in Lancashire for an industrialization that did not 'pauperize' independent artizans (Kitching 1982, Chapter 2). Neo-populist ideas, addressing the problems of the Third World and the failure of post-war growth strategies, emerged in the 1960s (Narda 1991). But:

> [neo-populism's] central moral concern remains unchanged, for it addresses the problem of inequality, of minority wealth and mass poverty.
>
> (Kitching 1982: 61, emphasis added)

As a political movement, neo-populism and populism is 'a collective awareness of disadvantage in relation to wealth' (Ionescu and Gellner 1969: 208), and in modern times has been the basis of Julius Nyerere's African Socialism (Nyerere 1962), Michael Lipton's Urban Bias (Lipton 1977), E.F. Schumacher's Appropriate Technology (Schumacher 1973), and much of the writing of the 'dependency' theorists (see the section 'Radical structuralism' below).

In the 1990s the micro-finance movement dealing with very small deposits and loans to micro-enterprises has been increasingly seen as an effective means to poverty reduction (Johnson and Rogaly 1997).

> The time has come to recognise microcredit as a powerful tool in the struggle to end poverty and economic dependence.
>
> (Daley-Harris 1997: iii)

The practice of micro-finance has two main problems: how to identify the poor entrepreneur in general, and the very poor entrepreneur in particular; and how to operate a credit system on very low rates. The poor (and/or the very poor) entrepreneurs are identified, and their activity assessed through techniques of impact assessment and impact monitoring. Impact assessment and monitoring is intended to identify changes in the social power-structure as a consequence of the economic activity of micro-entrepreneurs: how the first should become the last (Chambers 1997). How to build an awareness that the poorest are politically and often geographically remote, are weak, exploited and perhaps illiterate, are socially excluded, and are easily marginalized, and left out of empowering participatory processes. On this general theme – globalization, the poor, and entrepreneurial activity – see Montgomery *et al.* (1996) and Yaron *et al.* (1997).

With regard to interest charges, the question is one of how to manage the costs and risks of making financial services available to the poor. Here the experience of the Grameen Bank in Bangladesh is often cited as a model. Loans are made to people without any physical collateral, and through *peer group monitoring* (Jain 1996; Hulme and Moseley 1996a, 1996b), and loans are repaid in fifty weekly instalments, with 5 per cent of the loan repaid into a group fund.

Sustainable development, technology and institutions

> All of a sudden the phrase Sustainable Development (SD) has become pervasive. SD has become the watchword for international aid agencies, the jargon of development planners, the theme of conferences and learned papers, and the slogan of development and environmental activists . . . [but] 'What *is* SD?' is being asked increasingly frequently.
>
> (Lélé 1991: 607, emphasis in original)

For modernization theorists, where development is the outcome of billions of individual decisions, sustainability is a question of the number of consumers drawing upon finite natural resources. Pessimistic, egocentric theorists (see Chapter 7) only conceive of the issue of sustainability in terms of the geometric expansion of the human race compared with a finite natural resource base: population growth. At some point, inevitably, the carrying capacity of the planet will be exceeded, and people will start dying: either starving to death or suffering the effects of overcrowding and pollution.

However, optimistic egocentric theorists (see Chapter 7) have faith in individuals' ability to recognize their self-interest, and to respond to changing relative prices as the problems of sustainability become questions of short-ages and changing relative prices. But of course people will only be able to rationally so respond when they begin to think like independent individuals, and when society has modernized to adapt to individuals' personal prefer-ences. In Rostow's schema for modernization, there has to have been *take-off.*

... sustainability requires that all the world's population get past the second stage [the 'preconditions for take-off'] of the demographic transition.

(Baldwin 1995: 73)

Sustainability will 'naturally' evolve. We just have to be patient.

Structuralist theorists, however, are more proactive. On alternative conceptions of sustainable development, see Cole (1994). Sustainable development, 'development that meets the needs of the present without compromising the ability of future generations to meet their own needs' (WCED 1987: 43) is achievable only if the institutional basis of social and economic life, oriented towards fairness, is appropriate to the technical division of labour. The management of economic resources and the organization of social life must be relevant to the way in which we collectively survive. But the oft quoted definition of sustainable development cited above, from the World Commission on Environment and Development, often referred to as the Brundtland Commission after the Chairperson, Gro Harlem Brundtland the then Prime Minister of Norway, is too ambiguous to give any practical policy guidelines.

How should the present generation respect the needs of future generations? This has been answered in different ways, by different theorists, working within different schools of thought, within the structuralist perspective on development.

For Adrian Atkinson, the

actions of 'ordinary citizens' and 'community leaders' will only make informal tracks in the direction of insights of the ecophilosophers [to achieve sustainable development] via an adequate understanding of social and cultural structure and process and through conscious restructuring of this as a whole to produce a set of machinery that is inherently benign in its intercourse with nature.

(Atkinson 1991: 171).

Put simply, the problem is that the institutions of the world system, defined as a world technical division of labour, with the developing nations typically operating with relatively labour-intensive, inefficient, backward technology, do not reflect the social needs of the poor and disadvantaged. The world market, including stock exchanges, commodities markets, international banks, the IMF, the World Bank, and the World Trade Organization, serves to manage and maintain a system that acts to the benefit of the rich, industrialized, developed economies.

The economic, political and social life of developing economies, reflecting small-scale technology, calls for the 'recreation of the centre of social gravity at a very local level' (Atkinson 1991: 124). Institutions have to be decentralized, so that 'conscious decisions can be taken to actually control the general structure of social activity and its relationship with nature' (Atkinson 1991: 124). Such an analysis echoes E.F. Schumacher's call in 1974 for appropriate development: 'to bring help to those who need it most, each 'region' or 'district'. . .needs its *own* [small scale] development' (Schumacher 1973: 147, emphasis added).

It is a question of size. 'The question of scale is extremely crucial . . . in political, social and economic affairs' (Schumacher 1973: 55). People's social experience must be understood and grasped by the majority of the population. And the technical division of labour has become so specialized, and the resultant production process so sophisticated and complex, that it is beyond most people's comprehension; it has become 'more and more inhuman', and we need 'a technology with a human face' (Schumacher 1973: 122).

We need technology appropriate to the institutional scale of social life: *appropriate technology.*

For Schumacher and Atkinson, and for many structuralist theorists, including the ecocentrist environmental theorists addressed in Chapter 8, technology and production should be scaled down to the needs of people; people should not have to adapt to the needs of technology.

But *how?*

By convincing people that our life is unviable. But as yet there is little acknowledgement of the contradictions of modern life; there has not been a change 'in the form of consciousness with which our society apprehends life, no significant criticism of individualistic modes of thought and action' (Atkinson 1991: 216). This is a restatement of Hollis Chenery's 'failure of the equilibrating mechanism of the price system' above.

For E.F. Schumacher, change will come when mankind turns its back on the philosophy of materialism, 'which is now being challenged by events' (Schumacher 1973: 246). People must question 'conventional values', we have to 'free our imagination from bondage to the existing system' (HMSO 1972, quoted Schumacher 1973: 248), and in the pursuit of truth and justice we should be guided by the 'traditional wisdom of mankind' (Schumacher 1973: 250). It is not clear how we should recognise this wisdom when we find it, though for Schumacher it is revealed through Buddhist religious convictions.

However, in specifying the compatibility of social institutions and the technical division of labour, rather than production processes adapting to social, intellectual, moral and political habits, it can just as coherently be argued that economic progress has an evolutionary technical dynamic to which people *have* to adapt. Social institutions must change to meet the needs of the technical division of labour, not vice versa.

In this context, the World Commission on Environment and Development did not see the issue as one of discovering the traditional truths of social life:

[E]conomic growth is seen as the only way to tackle poverty, and hence achieve environment-development objectives [sustainability]. It must

however be a *new form of growth, sustainable, environmentally aware, egalitarian, integrating economic and social development.*
(WCED 1987: 59, emphasis added)

Such a 'sustainable, environmentally aware, egalitarian' way of life is not realized by people rediscovering traditional wisdom, but will evolve as people's experience highlights the technical conflicts and contradictions of social life. Technical evolution will beget social progress. The problem is essentially a rational one, to be solved by experts, trained to analyse the technical requirements of social existence: economists and development theorists, environmentalists, sociologists, planners, etc. And their conclusions and advice will become policy prescriptions if politicians have the courage to address difficult choices and make decisions that will harm some vested interests.

> We see 'ecological economics in one world' as a realistic utopia . . . solu-
> tions . . . could be realized given . . . *political will.* The stimulation of
> this will is the task of Green politics.
> (Group of Green Economists 1992: 11, 7, emphasis added)

Radical structuralism

Some structuralist theorists are not so sanguine about the possibilities of meaningful, institutional change: that is a change in social organization to effect the eradication, or at least the diminution, of poverty; and technical change to improve the security of people's livelihoods. They share neither the presumption of Hollis Chenery and the economists of the United Nations Conference on Trade and Development that such change is possible, nor the idealism of E. F. Schumacher and the economists of the German Green Party that a realistic utopia is more than a naïve illusion.

> Another group of writers – often referred to as the 'dependency school'
> – whilst usually accepting the structuralist analysis of the reasons for past
> failures of industrialization, takes a less optimistic view of the possibility
> of achieving policy change.
> (Colclough 1982: 496)

There have been two streams of dependency thinking: the *systemic* radical structuralists, and the *neo-Marxist* radical structuralists.

> An analytic scheme appropriate for the study of underdevelopment and
> for the formulation of strategies of development should be based on the
> concepts of process, structure and *system.*
> (Sunkel, quoted O'Brien 1975: 14,
> emphasis added)

For the systemic radical structuralists, the difference between a developed and an underdeveloped economy is that the former has an *endogenous growth capacity*: that is, the economy has an advanced technical division of labour, so that there is sufficient consumer demand to support the research and development needed for a producer goods (research and machine producing) sector. Underdeveloped economies lack this facility, and are dependent on the technology of other countries.

> [Dependence] is characterized by an absence of interdependence between economic functions of a system . . . the system has no *internal dynamic* which could enable it to function as an independent, autonomous entity.
>
> (Brewster, quoted Girvan 1973: 1, emphasis added)

This is an argument that evolved out of the United Nations Economic Commission for Latin America (ECLA) post-war analysis of Latin American economies.

Within the world division of labour there is inadequate linkage between low value-added primary production (agriculture and raw materials) and high value-added secondary production (manufacturing). These two functions are sited in different countries. As late as 1970 about 90 per cent of export earnings of the less developed countries came from primary production (Pearson 1969: 81). Hence the issue is a national one, rather than the Marxist emphasis on class. Modernization theory predicted the inevitable evolution of societies towards a modern high mass consumption utopia if trade was organized on comparative advantage: that is, countries specialize in those commodities that best utilize their resource endowment (where 'best' is defined in terms of world market prices and hence the value of resources). The centre exchanges industrial goods for the primary goods of the periphery.

> Technical progress in the centre would lead to lower prices for industrial exports, so that one unit of primary exports would eventually buy larger amounts of industrial imports – over the long term, progress would accrue to the periphery *without* industrialization.
>
> (Peet 1991: 44, emphasis added)

But in the 1930s and 1940s world trade slumped, consequent on the 1930s depression, and there were falling world primary commodity prices and declining export revenues. In Latin America, Argentina was particularly affected, as was Brazil, where the crisis was so serious that steam trains on the railroad burnt 'export' coffee beans as fuel. With low export incomes, imports were cut, and countries had to start producing imports locally: import substitution industrialization (ISI). The main theorist of the ECLA analysis was Raul Prebisch, formerly head of the Central Bank in Argentina.

The ECLA policy of ISI was not the ISI of Arthur Lewis addressed in Chapter 11; it was not intended to encourage market rationality, modernizing the minds of traditional producers. Rather, without structural changes to the economy, there could be no benign, market competition: the consequence could only be impoverishment, and technical development and trade would only advantage the industrialized countries. Prices would not fall in the advanced modern economies, as predicted by modernization theorists, because of the power of labour. Labour costs were continually raised by the threat of industrial action, and any gains in productivity from technical advances were more than compensated for by the higher wages paid to workers (Prebisch 1950). And in spite of technical progress the price of manufactured goods rose – an analysis subsequently developed by Arrighi Emmanuel (1971). Emmanuel argued that, because of the immobility of labour, wage levels are not equalized between countries, and hence the prices of traded goods, reflecting differential wage rates, mean that wealth is extracted from the low wage (and hence low price) economies to the advantage of the high wage (high price) economies.

> Peripheral countries export (agricultural) products that embody large quantities of their cheap labour and import (industrial) products embodying small amounts of expensive centre labour. This leads to terms of trade favouring the higher-cost products of the centre, while devaluing the exports of the periphery.
>
> (Peet 1991: 48)

And Samir Amin estimated that this mechanism of *surplus extraction* (transfer of value from the periphery to the centre) meant that an 'African peasant obtains . . . in return for 100 days of very hard work every year, a supply of imported manufactured goods the value of which amounts to barely twenty days of simple labour of European skilled work' (Amin 1976: 143).

This is unequal exchange: where value is defined as concrete labour time (not abstract labour time – see Chapter 5).

In addition to labour costs biasing export earnings, Prebisch pointed out that besides wage differentials, primary goods also have a low income elasticity of demand. That is, as incomes grow and people become richer, the demand for food does not grow at the same rate: the limits are physiological. And further, technical improvements to manufacturing processes reduce the raw materials content in manufactures, and synthetic substitutes are being developed for many primary goods: the limits are technical. All in all, the physiological and technical limits lower demand, leading to lower prices for primary commodities and hence reduced export earnings with which to purchase manufactured imports.

The only alternative was state, planned, industrialization. However, import substitution production was typically associated with high costs and poor quality, and as a policy strategy was only really applied in the late 1940s and early 1950s.

For the neo-Marxist radical structuralists the problems of dependency were not merely physiological, technical or organizational, but *political*. The state would not intervene in the economy to obviate dependence, because it was not in the interests of the ruling elite.

> The domestic elites in Latin America have formed a partnership with foreign corporations, and they would be most hurt by the economic reprisals of multinational corporations.
>
> (So 1990: 121)

Although it is the technical superiority of multinational corporations that bestows market and hence political power, the local political elite in dependent economies owe their privilege to being a link in the chain of exploitation, from the producers in peripheral, poor, *satellite* economies to central, rich, *metropole* economies. This 'extends the capitalist link between the capitalist world and national metropolises to the regional centres ... and from these to local centres, and so on to large landowners and merchants who expropriate surplus from small peasants or tenants ... Thus at each point, the international, national, and local capitalist system generates economic development for the few and underdevelopment for the many' (Frank 1969: 7–8).

> Latin America [is] in a situation of growing subjection and *economic dependence* ... in a single world system of ... commercial capitalism ... This ... relationship to the capitalist metropolis has formed and transformed the economic and *class structure* as well as the culture, of Latin American society ... [which] establishes very well defined class interests for the dominant sector of the [Latin American] bourgeoisie ... [producing] a *policy of underdevelopment* in the economic, social, and political life ... of Latin America.
>
> (Frank 1972: 13, emphasis in original)

This school of thought has been labelled neo-Marxist because certain words are used that evoke an impression of Marx's understanding of change as a consequence of class struggle (see Chapter 5). But neo-Marxist radical structuralism does *not* use Marxist concepts, and does not have a Marxist analysis of change and development; which is important in the way in which the *politics* of development is conceived.

> This concept [dependence] fails to grasp the real nature of the process of underdevelopment. The immediate explanation for this is the theoretical framework employed ... [which is] an eclectic combination of orthodox economic theory and revolutionary phraseology.
>
> (Kay 1975: 105)

Capitalism is not conceived of as a mode of production, but as a system of commerce. The emphasis is upon market exchange, and consequently it is argued that capitalism has existed for about 5000 years.

> ...the same continuing world system, including its centre–periphery structure, hegemony–rivalry competition, and cyclical ups and downs has been evolving (developing?) for five thousand years at least ... In this context, the mixtures and variations of different 'modes' of production ... are much less important than the constancy and continuity of the world system ...
>
> (Frank 1996: 40–1)

However, for Ernesto Laclau it is the particular contradictions *within* modes of production – for instance the feudal mode of production, leading to unequal exchange between Europe and Latin America and 'reducing the economic surplus of the peripheral countries and fixing their relations of production in an archaic mould of *extra-economic* coercion' (Laclau 1971: 35, emphasis added) – that are important. So that, as the capitalist mode of production emerged in Europe, where class power is a consequence of economic (not extra-economic, political) coercion through market exchange – commodity exchange, see Chapter 5 – and with the inherent tendency for the rate of profit to fall, the pre-capitalist character of underdeveloped economies 'maintained the average rate of profit by offering opportunities for higher rates of return' (Leys 1996: 62), not based upon the competitive dynamic of market exchange. The theoretical emphasis is on the conditions of production, not trade and exchange, with distinct implications for strategies to ameliorate poverty.

Neo-Marxist radical structuralism built on Paul Baran's classic 1957 study *The Political Economy of Growth* (Baran 1967). For Baran, competitive world trade, in contradistinction to the modernization theorists, did not encourage development, but *under*development. The price structure of the system of world trade extracts economic surpluses from the periphery to the advantage of the metropole. As noted above, exploitation is a consequence of trade, and not a result of the class nature of the control of the means of production. And hence the development of the metropole and the underdevelopment of the periphery are systematically contiguous in the world economy: the former is a consequence of the latter. Trade and monopoly power develop the exploitative link that extends

> in chain like fashion ... between the ... national metropoles to the regional centres ... and from these to local centres, and so on to the large landowners or merchants who expropriate surplus from small peasants or tenants, and sometimes even from these latter to landless labourers exploited by them in turn. At each step on the way, the relatively few capitalists above exercise *monopoly power* over the many below ...
>
> (Peet 1991: 46–7, emphasis added)

Development within the parameters of international exchange is 'impossible within the periphery' (Peet 1991: 52). And as such the analysis of underdevelopment calls for a policy of *autarchy*. Peripheral societies, as far as possible, should be self-sufficient, and make a radical break with the world market, pre-empting the monopoly power of developed economies.

> [Developing societies should] delink from the system externally and . . . adopt self-reliant socialism . . . internally in order to make independent or nondependent economic development possible.
>
> (Frank 1996: 28)

But how socialist self-reliance is to be attained is not theorized. There are few policy suggestions as to what marginalized, impoverished people living in the periphery can actually *do* to relieve their plight. World systems theory closely follows the theoretical logic of neo-Marxist radical structuralism (Peet 1991: 49–51), but with a more 'functionalist' emphasis, allowing even less scope for the disadvantaged to actively participate in their well-being. The World System is

> a single *division of labour*, comprising multiple cultural systems, multiple political entities and even different modes of surplus appropriation.
>
> (Wallerstein 1980: 5, emphasis added)

The system, based on a shared technical division of labour, makes all the parties to world trade dependent on each other. In contradistinction to dependency theory, economic development requires trade: autarchy and self-reliance is not an option for development. Like the neo-Marxist radical structuralists, for world systems theorists market monopoly power lies with whosoever controls the technology that underlies the technical division of labour.

And now not only is the world divided between satellite (periphery) and metropole (core) economies and societies, but there is also an intermediate group of semi-peripheral countries between the periphery and the core. In a very functionalist sense, generating a fatalistic apathy towards relieving poverty, these are a *necessary* buffer, with intermediate wages and intermediate profits, which allows for the upward and downward movement of nations between peripheral and core status. And this buffer allows capital to discipline labour in the core, tempering wage demands and improving profitability.

The radical structuralist/dependency/world systems emphasis helped to underpin a strong populist tradition 'favouring domestic policies of economic nationalism, of self-reliance and of delinking [from international markets]' (Hoogvelt 1997: 42). The issue is one of capitalist industrialization, and the non-capitalist elites of the underdeveloped economies, unable to maintain their advantage and privilege in an increasingly capitalist world, through international negotiating forums, conferences, and organizations, called for fundamental change in the world economy: higher and more secure prices; preferential access to the domestic markets of the developed economies for their traditional exports and for the exports of their emergent infant industries; reform of the international monetary system; more economic aid and technical assistance; codes of conduct for multinational corporations; etc. – a tide of opinion that in 1974–75 led to a United Nations Charter of Economic Rights and Duties of States, and a Programme of Action for the Establishment of a New International Economic Order (NIEO) (Willetts 1978).

There was to have been institutional change in the world economy by which experts, for example the economists of the United Nations Conference on Trade and Development (UNCTAD), would purposefully reorganize the relations of technical dependency between rich and poor societies. The ideology of the NIEO (and radical structuralism in general) emphasizes the activity of 'planners in the state apparatuses of Third World countries, of their advisers ... in regional policy bureaux such as ECLA, and of the personnel of the "aid and trade" network, from the UN Conference on Trade and Development (UNCTAD) [etc.] ... [and] progressive intellectuals (including academics)' (Leys 1996: 63).

And in this sense it is very elitist.

However, it was never clear how management by a benign elite would be achieved, and anyway, changes in the world economy were so rapid as to make the NIEO anachronistic. The NIEO was essentially a strategy by which different countries could manage relations of exchange. But the emergence of transnational capital in the 1970s and 1980s, where different parts of the same production process are located in different countries, made competition between nation states an irrelevance.

> [R]elations of capitalist domination and exploitation ... [have to be] conceptualized in terms of global class relations which transcend national class structures.
>
> (Hoogvelt 1997: 58, emphasis added)

Industrial development in developed and underdeveloped economies became increasingly a question of the political authorities' guaranteeing a cheap, compliant, and sufficiently educated workforce, and offering tax concessions to international investors, as well as the minimum of regulation: the sort of environment that the Multilateral Agreement on Investment (MAI), addressed in Chapter 11, is trying to guarantee globally.

And for Gunder Frank, dependency is not now a question of the control of industrial production and the international division of labour, but an issue of international debt:

> [D]ebt is an instrument of neocolonization and a drain of 'surplus' from part of the South. By my calculation, this loss of capital from South to North has been of the order of US $100 billion per year ... The result is ... underdevelopment. This time it is with *dis*investment in

productive infrastructure and human capital and with a loss of competitiveness on the world market . . .

(Frank 1996: 34–5, emphasis in original)

The apparent success of the East Asian Newly Industrialized Countries (NICs) up until the mid-1990s prompted a rethink of the structuralist fatalism of dependency theory: a theory that predicted that underdeveloped economies could only get poorer, and conceptualized marginalized people as passive victims. Recognizing the role of transnational capital, and that it is clearly possible to break out of the cycle of underdevelopment, Frank has now accepted the theoretical analysis of World Systems Theory: that trade is an inevitable aspect of development. The world economy is still determinant in underdevelopment, but existing models of development are unable to address and ameliorate the effect of the world market on dependent peoples:

> This inadequacy characterizes the magic of the world and domestic market, Western top-down political democracy, Eastern top-down economic democracy and recent attempts at self-reliant national state delinking . . . Nor does anything on the horizon offer most of the population in much of the Third World any chance or hope of equity or efficiency in economic development . . . [But] A luta cotinua! – the struggle continues.
>
> (Frank 1996: 46–7, 52)

The struggle might be continuing: but by who, and following what strategy? What should be the priorities in defining activity and policy formulations aimed at relieving poverty and exploitation? World systems/dependency theorists appear to be as unsure and confused as ever, relying on a casual, empiricist notion of the real world, and an eclectic combination of populist theoretical conceptions in the name of pragmatism.

> . . . dependency theory's focus . . . remains *indisputably* valid . . . dependency theory has passed into the standard conceptual toolkit of most people who *seriously study* problems of development.
>
> (Leys 1996: 31, emphasis added)

But are countries poor because their economies are dependent; or is economic dependency a consequence of being poor? And in the case of the latter, what creates poverty?

But this is a question outside the intellectual parameters of Colin Leys' understanding.

Those theorists who dispute the determinist, technocratic, elitist logic of dependency theory are clearly not serious students of development, and can be dismissed as not looking closely enough to see the real world.

The 'impasse'

We saw in Chapter 11 that modernization theorists, trying to explain the apparent failure of free market economic policy to ameliorate the predicament of the poor, are looking to new institutional economics to explain why individuals are unable to act in their own best interests, because of institutional constraints.

Similarly, the management strategies of structuralist theorists have not unambiguously been the harbinger of progress. And although Gunder Frank puts his faith in an unspecified struggle, and Colin Leys feels that the constraints on development as a result of dependency are indisputable, some structuralist theorists think that the concept of development has to be fundamentally rethought.

> Many developing countries will remember the 1980s as the lost decade. The same assessment could perhaps be applied to the field of development theory. Especially from the mid-1980s onwards, an increasing number of publications outlined the contours of what became to be known as 'the impasse in development theory'.
>
> (Schuurman 1993: 1)

A structuralist, management strategy for development has to be based on a 'model' of the development process, to identify and highlight how progressive change takes place. ('Progressive' here is defined as a greater technical facility to control and productively use the natural environment, and a greater 'fairness' in the distribution of production.) By modelling development processes, the significant factors in the development dynamic can potentially be identified and acted upon through policy prescriptions. As the world technical division of labour changes and evolves, and new institutions arise to manage and coordinate social activity to effect production, then, if the world economy is to be efficiently and equitably managed, new models, or *paradigms*, have to be conceptualized and developed.

In particular, development theory has to be globalized.

> . . . the world market is an over-arching whole which cannot be approached using development policies oriented at the national level. Individual nation-states are assigned an increasingly smaller function. Development theories, however, still used the nation state as a meaningful context . . .
>
> (Schuurman 1993: 10)

Such a model, or paradigm, would account for the inequities and lost technical potential within the world system, and offer a realistic development management strategy, which through institutional change and expert management would allow us all to share in the unprecedented ability of humans to technically control the natural environment.

The search for this new paradigm is the theoretical impasse of structuralist development theory. Because theorists essentially are confused over the *dynamic* of social and economic change, 'many writers have turned back to reconsider the essential nature of "human agency"' (Long 1992a: 22).

How do we understand people's attitudes, intuitions and social behaviour? How do we interpret 'human agency'?

> The essence of the actor-oriented approach is that its concepts are grounded in the everyday life experience and understandings of men and women.
>
> (Long 1992b: 5)

However, for structuralist theorists, people's behaviour is not simply a reflection of individuals' particular subjective preferences and personal interests. Social life is more than the lives of individuals in society. To understand their own lives, and in the search for order and meaning to effect an element of control over their futures, 'the individual is . . . transmuted metaphorically into the *social actor*' (Long 1992a: 25, emphasis in original). So, how do we begin to define and identify the social context of individuals' behaviour: the *social* actor?

> We should be careful . . . to restrict our use of the term 'social actor' . . . to those social entities that can meaningfully be attributed with the power of agency . . . knowledge processes are embedded in social processes that imply aspects of power, authority and legitimation.
>
> (Long 1992a: 23, 27)

In a parallel attempt to return to the 'power of human agency', and to find an alternative to the determinist structuralist theories of social change, which had been thought to explain social power and the resultant trajectory of social change and development, David Booth (1985), rejected 'meta-theories', explaining the necessity of development (and underdevelopment), leaving little role for Norman Long's social actor to purposefully intervene to alter the course of events.

We have, then, to explain people's differing responses and motivations to understand social trends.

> New or recently revived theoretical influences have combined with changes in the world to generate a fresh intellectual climate, typified by an enhanced interest in the *diversity* of development experience across different national, regional and local settings.
>
> (Booth 1994: 298, emphasis added)

In particular, research programmes into development processes need to be designed to identify 'significant regularities or *patterns* of difference' (Booth

1994: 306, emphasis in original). But phenomena are not *abstractly* diverse or different; they have to be defined as different in some particular way, with reference to a particular standard of normality. If development experience is 'diverse', it must be diverse, at least implicitly, with regard to some conception of what development *is* or *should be*. We have to have a theory of development to be able to highlight the 'illumination of difference' (Booth 1994: 310).

Without a theory of development, without a theory of *why* people change their behaviour to effect an improvement in living standards or ways of life, phenomena and processes we identify as 'different' reflect the values, priorities and intuitions of the researcher. The only world that can ever be known is a world of difference, and progress and development beyond the bias of the researcher, cannot be addressed. There is a naïve belief that the facts speak for themselves.

> ... new theories ... will materialize only on the basis of fresh *empirically* rooted analytical *comparisons* ... it will be more or less concerned with ... *dominant* socio-economic institutions ... [and] involve a major element of inductive generalization from new research.
>
> (Booth 1994: 307, emphasis added)

And 'inductive generalization' implies an incipient theory of what is to be explained. *Which* data are 'empirically' relevant can only reflect theoretical emphases and priorities, and 'comparisons' can only be made by reference to a wider conception of development, suggesting which socio-economic institutions are considered to be dominant, and why. For David Booth

> new development theory will emerge autogenetically from the accumulating volume and density of all this [research] work, through some *spontaneous* fusion ... [but there are] implicit higher-level theoretical presuppositions that need to be made explicit ... And even more crucially, the construction of a new theory of development is necessarily a political task, involving *political choices* about whom (what social forces) the theory is for ...
>
> (Leys 1996: 28, emphasis added)

Choices have to be made based upon a theoretical understanding of the development process.

13 Development as the fulfilment of people's social potentials

Class struggle

People: passive or active?

For modernization theorists, development is a consequence of the inevitable need of independent individuals to assert their unique preferences – a biological imperative to be 'free' – albeit that this freedom can only be generalized to the population as a whole if the state is proactive in guaranteeing the rights of all individuals to act according to their subjective preferences. To this end societies have to be modernized to engender development: the *modernization of the mind*.

Development is a question of individuals' attitudes, and how individuals think.

> Failure to understand that the roots of economic behaviour [and hence development] lie in realm of consciousness and culture leads to the common mistake of attributing material causes to phenomena that are essentially *ideal in nature*.'
>
> (Fukuyama 1992: 7, emphasis added)

But for structuralist theorists people are not free to choose, free to make up their minds and act in *their* best interest. Humans are not independent individuals; they are dependent on each other. Social life is not a consequence of people competing to enjoy themselves (maximize utility), but of cooperating within a technically defined division of labour in production.

With evolving technology, structural constraints impede efficient technical cooperation. Individuals are organized to work cooperatively through an institutional structure, which not only technically coordinates productive activity, but also distributes the product of such coordination according to culturally defined standards of fairness. Without broad agreement within society to such standards cooperation would be stymied by conflict over the distribution of the technical product, and ultimately society would collapse. And it is the intellectual project of structuralist development theorists to specify the institutional prerequisites for a viable society, given the extant technological parameters of the division of labour in production: prerequisites that have to take into consideration the historical experience of institutional change within society.

Of course, as we have seen, the radical structuralists, especially the neo-Marxist radical structuralists, understand such processes of institutional change to be inhibited by vested economic interests, which control the political, institutional structure of social organization.

Where individuals are believed to be essentially either independent or dependent beings, social change (and hence development) is more or less determined and beyond people's control. All that can be achieved is to organize social life to take advantage of the biological or technical parameters within which people are believed to exist. Individuals' behaviour is either a consequence of their genetic inheritance, or the result of the inexorable technical cooperation between people within a technical division of labour. It is the role of the development theorist to define the room for manoeuvre either within biological parameters or within technical limits.

A third perspective on development does not see individuals as passive in the face of biological or social forces beyond their control. Rather people are *active* participants in the process of development. Certainly people have a genetic inheritance: particular biological characteristics that are innate. But these characteristics can only become potentials through social experience. And people's social experience will reflect their potentials: social life is an emergent property of independent individuals depending on each other: people are *interdependent*.

The dialectics of human nature

An attribute cannot become a talent without the catalyst of social experience. As for the abstract labour theory of value in economics (see Chapter 5), where the consumer and the producer are dialectically related in the citizen as economic dynamic, so in the development process people's potentials can only be understood as a dialectic of their biological endowment *and* their social experience.

In Chapter 5 it was argued that through *praxis* – the process by which awareness and consciousness are changed as a consequence of experience – people are in a process of constant change (see also Chapter 17). And when people are frustrated and constrained from realizing their potentials, then, if they are aware of, and organize and combine with, others who are similarly constrained, people who share the same class interest, then purposeful change towards fulfilling human potentials can be effected.

Purposeful change, at least in part, reflects class-consciousness – the awareness of people of a common class interest *between* themselves. Individuals do not 'possess' class-consciousness; it is a process of cooperation between people. People's everyday frustrations – unemployment, high prices and/or low wages, inadequate health care or education facilities, etc. – are not obviously class issues. To be conscious that individuals' distinct problems and constraints may systematically be linked, may be different manifestations of a common cause – the relative powerlessness of those who do not control the means of production (see Chapter 5) – requires the abstract, theoretical interpretation of diverse experiences to reveal shared social constraints. The appearance of social life disguises a shared essence of social existence, which it is the purpose of theoretical analysis to reveal.

> . . . theory . . . becomes a material force once it seizes the masses. Theory is capable of seizing the masses once it demonstrates *ad hominem*, and it demonstrates *ad hominem* once it becomes radical. To be radical is to grasp matters by the root. But for man the root is man himself.
>
> (Marx 1970: 137)

However, the disadvantaged are only relatively powerless. Power is a relationship: it is two-way. It is not a 'zero-sum' game: something that the powerful possess and the powerless do not. Those in authority in general, and capitalists in particular, certainly are able to deny people access to resources or livelihoods. But those people who are able to allocate resources and incomes also need something to allocate. Capitalist producers need wage labour as employees. As we saw in Chapter 5, value and surplus value (and hence profit, rent and interest) are a consequence of the employment and exploitation of labour power. And if the disadvantaged, proletarianized majority organize and combine to oppose the power of the privileged, bourgeois minority, who organize production for the purpose of commodity exchange and the realization of surplus value, leading to economic crisis, then potentially the powerful can be challenged and change effected. But, of course, the state as the guarantor of individual freedom militates against the freedom of the disadvantaged and exploited to organize and prosecute their class interest.

> What crushes the element of real choice out of the lives of isolated individuals is their total separation from any . . . [organized] movement and their utter dependence, both economically and ideologically, on a system that is entirely hostile to their needs and aspirations. The more isolated and powerless the individual and the more brutal the circumstances he or she confronts the less chance he or she has of influencing their individual fate.
>
> (Rees 1998: 283)

This isolation is compounded by capitalist relations of production of domination and exploitation becoming ever more international. To challenge ruling class power means to address capitalist relations of production on a world scale. It is not enough to identify with the nurse who treats your mother, or the lorry driver you play football with, or the teacher of your child: Colombian men and women who work for British Petroleum, or Nigerian men and women who work for Shell oil, are equally your brothers and sisters in class struggle, as is anybody who suffers the rapacities of inequality as a consequence of being a wage labourer.

When Hitler attacked the Jews I was not a Jew, therefore I was not concerned. And when Hitler attacked the Catholics, I was not a Catholic and therefore I was not concerned. And when Hitler attacked the unions and the industrialists, I was not a member of the unions and I was not concerned. Then Hitler attacked me and the Protestant church – and there was nobody left to be concerned.

(Martin Niemoller, quoted MacArthur 1998: inside page)

'Trotsky's theory of the law of combined and uneven development stressed that any analysis of the revolutionary potentiality of backward countries must start from the *totality* of capitalist development on a world scale' (Rees 1998: 283, emphasis in original). Development is inevitably combined through commodity exchange (see Chapter 5), but uneven (and unequal) as a consequence of class power (Smith 1990, Chapter 6).

However, an awareness of the internationalism of class struggle is not to posit the probability, or even possibility, of world revolution, in the sense of the exploited and disadvantaged of the world simultaneously rising up to challenge world, bourgeois power. Individuals are motivated to fight for social change when they are frustrated in their daily life from fulfilling their potentials. And even the awareness of what their potentials might be is itself a reflection of their daily experience and their conscious appreciation of that activity. Consequently social change and development is inevitably a step-by-step process of trying to make tomorrow a little better than yesterday: in the *present*, imagining the *future* possibilities implied by *past* conflicts. Change is rarely explicitly motivated to try to achieve an utopian ideal.

People struggle over issues – housing, the environment, armaments, hunger, repression, fair trade, etc. – and for these protests, referred to in the literature as 'social movements', to be coalesced into a class movement challenging economic power is the role of the political activist, the revolutionary intellectual, working with and through a political party.

Class consciousness is a social construct which, however, does not make it less 'real' and important in history. While the social forces and expressions of class consciousness vary, it is a recurring phenomenon throughout history and most of the world, even as it is overshadowed by other forms of consciousness at different moments (that is, race, gender, national) or combined with them (nationalism and class consciousness).

(Petras 1998: 32)

However, politics is more than this, and cannot simply be conceived in terms of processes within institutions and nation states. With the worldwide scope of the capitalist mode of production and the international allocation of resources and control of economic activity, politics and the political party, the organization of disparate individuals to prosecute a common class interest has to be understood in a cultural context. While national governments, and political parties and institutions, are crucial to organizing daily social and economic activity, 'the policies and practices of states in distributing power upward to the international level and downwards to sub-national agencies are the sutures that will hold the system of *governance* together' (Hirst and Thompson 1996: 184, emphasis added), where 'governance' refers to the culture of the social control of individuals' activity oriented to the achievement of a range of desired outcomes. This process is not merely the purview of the state and political agencies.

As a cultural process, governance evolves with social and political change as the need to address people's changing and differing potentials becomes ever more an explicit, political priority with people's increased class-consciousness. It is not a question of whether 'a coherent *system* [of governance] will develop' (Hirst and Thompson 1996: 184, emphasis added), but how far people's intuitions and motivations consciously reflect the process of social change: praxis (see Chapter 17). In the development process, people increasingly empower themselves to participate in the social realization of their individual creative potentials.

Communism

> One major aim of Marx's analysis of capitalism is to explain how people can make their *own* history and be made by it at the same time, how we are *both free and conditioned*, and how the future is both open and necessary.
>
> (Ollman 1993: 89, emphasis added)

In as far as people are aware of their potentials, the future is open; in as far as the exigencies of social existence require cooperation to fulfil these potentials, the future is necessary. For Marx the potential was *communism*. But this conception of social organization was not conceived of as a blueprint for social organization, only what, logically, *could* be possible, if everyone shared a consciousness of class interest, in the context of people's *changing* creative potentials. Communism is not to be imposed by the institutions of the state; it evolves as a consequence of free activity.

> Only in community [with others has each] individual the means of *cultivating his gifts* in all directions; only in community, therefore, is personal freedom possible . . . In a real community the individuals obtain their freedom in and through their association . . . in communist society, where nobody has one exclusive sphere of activity but each can become accomplished *in any branch he wishes*, society regulates the general production and thus makes it possible for me to do one thing today and another tomorrow, to hunt in the morning, fish in the afternoon, rear cattle in the evening, criticism after dinner . . . without ever becoming hunter, fisherman, herdsman or critic.
>
> (Marx and Engels 1970: 83, 54, emphasis added)

For Marx and Engels, individuals' imperative to fulfil their potentials (see Chapter 17) through praxis, changes people's intuitions and nature: society increasingly becomes class-conscious.

> . . . the communal production by society as a whole . . . will both require and generate an entirely different kind of human material. Communal operation of production cannot be carried out by people as they are today . . . Communal planned industry operated by society as a whole presupposes human beings with many-sided talents and the capacity to oversee the system of production in its entirety . . . Society organised on a communist basis will thus give its members the opportunity to put their many-sided talents to many-sided use.
>
> (Engels 1977: 19–20)

Communism, communal production by society, is emergent:

> . . . communist society . . . *emerges* from capitalist society . . . [and] is thus in every respect, economically, morally and intellectually, still stamped with the birth marks of the old society from whose womb it emerges.
>
> (Marx 1972b: 15, emphasis in original)

Progress, the trend towards communism, 'socialism', is the emergence of class-consciousness in world society.

> We make and change the world only through the mind of man, through his will for work, his longing for happiness, in brief, through his psychological existence. The 'Marxists' who denigrated [sic] into 'economists' forgot this a long time ago.
>
> (Gramsci, quoted Boggs 1976: 57)

The Marxist economists posit a technological determinist interpretation of the dialectic of social change, but for Marx what is distinctive about human beings and social change is their self-consciousness, and their creative potentials (see Chapter 17). People are aware of themselves, and can choose how to best fulfil their potentials, potentials that change with social experience: and through their changing behaviour society changes. But change is only effected by challenging the interests of the ruling class. And progress is a consequence of class-consciousness: that is, change that has, often as a by-product or an unintended consequence of prosecuting a more immediate issue, sufficiently challenged power to allow the disadvantaged more scope to fulfil their potentials.

Socialist development

All human beings are intellectuals: all have beliefs, ideas, feelings, intuitions, etc. And all people make choices as to how better to organize their lives. Though of course in an unequal society not all people equally have the opportunity to develop the mental and psychological potentials of their being. Additionally the poor and the disadvantaged have fewer options available to affect their lives. This question of consciousness and the understanding of the human condition is treated in detail in Chapter 17, and will not be previewed now. The concern here is to focus on the process of *socialist development*.

> Socialism is . . . a process of successive upheavals not only in the economy, politics and ideology but in conscious and organized action. It is a process premised on unleashing the *power of the people,* who learn how to change themselves along with their circumstances. Revolutions within the revolution demand creativity and unity with respect to principles and organization and broad and *growing* participation. In other words, they must become a gigantic school through which people learn to direct social processes. *Socialism is not constructed spontaneously, nor is it something that can be bestowed.*
>
> (Heredia 1993: 64, emphasis added).

People are only free to be themselves, to fulfil their potentials, when they have fulfilled the material requirements for daily life: food, clothing, housing, and all the associated social and political prerequisites, education, culture, democracy, etc., which are part and parcel of material production.

> In fact the realm of freedom actually begins where labour which is determined by necessity and mundane considerations ceases; thus in the very nature of things it lies beyond the sphere of *actual material production.*
>
> (Marx 1972a: 820, emphasis added)

A socialist development strategy implies a society based on participatory, political democracy, and an economic democracy geared to the provision of social need, not private profit. And the emerging participatory political and economic democracy is a consequence of a developing class-consciousness, an awareness of the self in society, predicated on understanding the social significance of individuals' intuitions. Hence all projects must be based on community agreement between civic, religious and political groups and interests – an effect of *civil society.*

> Development and education are first of all about liberating people from all that holds them back from a full human life. Ultimately development and education are about transforming society . . . Development,

liberation and transformation are all aspects of the same process. It is not a marginal activity. It is at the core of all *creative* human living.

(Hope and Trimmel 1995: 9, emphasis added)

Development theorists become political activists, facilitating people's understanding of the causes of their disadvantage; developing practical community alternatives to improve people's well-being – literacy programmes, health-care provision, the social coordination of community life, productive activity to satisfy community needs, etc.; and crucially, throughout all of these development activities, building confidence that change is possible, encouraging creativity and dispelling any fatalistic acceptance by people of the status quo. There is an emergent class-consciousness. The institutions of daily life have to be transformed to operate by democratic principles, with officials accountable to the community, and activity must be analysed to ensure that those in authority address people's changing needs.

Such a development policy depends upon: improved and continually improving communication, listening to others; the conceptualization and articulation of people's intuitions and insights; the diagnosis of needs and the analysis of problems; and people planning and acting together communally in teams and organizations.

Those with education and skills have a role to play in enabling the poor to participate actively in identifying and analysing the causes of their problems, uniting with them in finding solutions. The needs of the people must be strategically linked to public policy. Coalitions of organisations and community groups must be built to advocate these policies.

(Hope and Trimmel 1995: 12)

Socialist development: the case of Cuba

In the argument so far, socialist development has been discussed in very general, theoretical terms. But the purpose of theory is to understand experience, not supplant it: to be conscious of the social consequences of our individual intuitions so that we can orchestrate and direct our activity to fulfilling our potentials within society. And as we more fully understand other people, so we know ourselves, and are able to know others better: the dialectic of the individual in society.

A dialectical approach to the understanding of social change and development can be applied to the experience of *any* society, the practice of social life being understood to be a consequence of contradictions between opposing class interests – even though people are almost invariably unaware of the essential class basis of their actual activity. The concept of class does not describe experience; it *explains* it. Development, progress towards the better fulfilment of citizens' potentials – the evolution towards communism, or the process of socialism – is a property of social life, even within societies that do not have an explicitly socialist development strategy, such as Britain in the 1990s. Even though socialism is not explicitly on the political agenda, in the late twentieth century people in Britain are able to fulfil far more of their potentials than in the seventeenth, eighteenth or nineteenth centuries. And this empowerment has not merely been a consequence of technical improvements in productive efficiency, though this has made such change possible, but has been achieved by an ongoing process of class struggle. See Cole and Postgate (1961), Thompson (1968), Baker (1975) and Abendroth (1972).

Socialism is a process, not the specification of a set of institutions intended to create equality: a process of social change consequent on the inevitable struggle of individuals to achieve their social potentials.

However, with regard to understanding an explicitly socialist development strategy – the process by which people empower themselves in the attempt to fulfil their potentials – it is instructive to look at the experience of Cuba.

> The Cuban revolution declared, from the outset, that no one should go malnourished. No disappointment in food production, no failed economic take-off, no shock wave from the world economic crisis has deterred Cuba from freeing itself from the suffering and shame of a single wasted child or an elderly person ignominiously subsisting on petfood. No other country in this hemisphere, including the United States, can make this claim.
>
> (Benjamin *et al.* 1994: 189)

The United Nations Development Programme publishes an annual *Human Development Report*, which looks at the record of poverty alleviation in 129 developing countries, from Afghanistan to Zimbabwe, in terms of a human development index (HDI) and a human poverty index (HPI). The HDI is based on three indicators: longevity as measured by life expectancy at birth; educational attainment, combining a measure of adult literacy, and primary, secondary and tertiary enrolment ratios; and a standard of living index based on an estimate of gross domestic product (GDP) per capita (UNDP1997: 122). And the HPI highlights the scale of human deprivation in each of these fundamental aspects of human existence: the percentage of people not expected to survive to age 40; the deprivation of knowledge, and hence the level of illiteracy and enrolment; the standard of living as measured by access to safe water, health services, and underweight children (UNDP 1997: 125).

> According to the Food and Agricultural Organization of the United Nations (FAO), and the United Nations Children's Fund (UNICEF), 800,000 children that could be saved, die every year in Latin America, and none of those children die in Cuba ... Thirty million children are homeless, in Cuba there are none; there are tens of millions of beggars, there are none in Cuba.
>
> (Castro 1991: 51, my translation)

As measured by the HDI and HPI indexes, Cuba is in the top group of five developing countries; and this in spite of the privations suffered as a result of the collapse of the Soviet bloc and the Communist trading organization CMEA (Council for Mutual Economic Assistance) within which Cuba conducted some 80 per cent of its international trade, *and* the US-sponsored economic blockade, first formalized in 1963 although initiated earlier, and tightened by the Cuba Democracy Act (Torricelli Bill) in 1992.

> Not until 1993 ... did Cuba face the full and combined force of the collapse of Soviet-bloc Communism and the US economic blockade ... In 1993 primarily because of this 'double blockade' and particularly destructive weather conditions ... trade collapsed. The effect was dramatic. Hospital equipment without spare parts goes unrepaired. Medical doctors, lacking medicines, anxiously seek herbal cures. Newspapers and magazines are in short supply: no paper. Lack of gas, batteries, and tires cripples buses, trucks and cars. Stores, offices, and homes darken as electrical output falters. Cooking gas is available for only a few hours each morning and evening.
>
> (Fitzgerald 1994: 174, 1).

This predicament was exacerbated in March 1996, when President Clinton endorsed the harshest ever package of measures against Cuba by approving the Cuban Liberty and Solidarity Act (Helms-Burton Act) (Cole 1998, Chapter 1).

The double blockade saw the global social product of the Cuban economy (equivalent to gross national product, or national income) plummet by 25 per cent in 1991; but by 1994 there was positive growth of 0.7 per cent, 2.5 per cent in 1995, 7.8 per cent in 1996, and 6 per cent in 1997.

Cubanidad

The Cuban revolution did not start out as an explicitly socialist revolution; it adopted the socialist nomenclature only after the failed US-sponsored invasion in 1961 at the Bay of Pigs. And even then, on 17 August, four months after the invasion, Che Guevara, who was in Uruguay, told Richard Goodwin, US President Kennedy's Assistant Special Council, that Cuba would forgo an alliance with the Soviet Union, compensate US owners of confiscated property, and curb support for left-wing insurgents in third world countries if the United States would cease hostile action (*Guardian* 30 April 1996).

Since the victory of Fidel Castro's 26 July Movement over the forces of Fulgencio Batista Zaldívar on 1 January 1959, two themes run through the history of the Cuban revolution: the political imperative for people to fulfil their potentials – 'Our task is to enlarge democracy within the revolution as far as possible ... to assure channels for the expression of the popular will' (Che Guevara in 1961, quoted Zeitlin 1970: 78); and the economic imperative for the Cuban economy to escape dependency on sugar production – 'Cuban development strategy since the Revolution has always been in the context of a dependent external sector which has been a major obstacle' (José Luis Rodríguez, Minister of the Economy and Planning, 1990: 209, my translation).

Efforts to realize people's potentials in the context of economic dependency explains the process of Cuban socialism.

Cuba has never been a Caribbean outpost of the Soviet Empire, and even before the end of the Cold War and the collapse of the East European Communist bloc in 1989–91 the Cuban revolution has always been infused with 'Cubaninity': *Cubanidad*.

> Capitalism sacrifices the human being, communism with its totalitarian conceptions sacrifices human rights. We agree neither with one nor the other ... Our revolution is not red but olive green. It bears the colour of the rebel from the Sierra Maestra.
>
> (Fidel Castro, 21 May 1959, quoted Binns and González 1980: 13)

The political project of the Cuban revolution in 1959 was to achieve self-determination, continuing the revolutionary struggle initiated by José Martí in the nineteenth century against Spanish colonialism, which was suffused by American domination when the United States entered the War of Independence against Spain, a war that ended in 1898. In 1901 the American government imposed the Platt Amendment on Cuba, giving the United States the right to restrict Cuban sovereignty in international affairs, and the 1903 Reciprocity Treaty tied Cuba's trade into a dependent relationship with the USA (Pérez 1982: 192–9). The Cuban economy had been effectively 'confiscated' by the United States (Blackburn 1963).

> [We] made politics our only industry and administrative fraud the only course open to wealth for our compatriots ... This political industry ... is stronger than the sugar industry, which is no longer ours; more lucrative than the railroads which are managed by foreigners; safer than the banks, than maritime transportation and commercial trade, which also do not belong to us.
>
> (De Carrión, quoted Pérez 1982: 215)

The development strategy implied by the ethics of the revolution quickly led to the nationalization of economic activity, the expansion of welfare provision and the delivery of health services and education as of right by the state, which meant an enormously enlarged public sector. And the United States was not slow to respond to the nationalization of the property of US corporations and citizens, even though compensation was offered, and accepted by other countries similarly affected. Retaliation took the form of an embargo, which lasts to the present day (November 1998), numerous assassination attempts on Fidel Castro, including the planting of exploding cigars by the US Central Intelligence Agency, and military action, including the (defeated) US invasion at the Bay of Pigs in 1961.

At the time of the revolution there was no economic development strategy, other than to reduce foreign dependence, and to make the economy more self-sufficient through import substitution. To this end JUCEPLAN (Junta Central de Planificación), the Central Planning Board, was created to plan and coordinate economic activity. However, the lack of an accurate statistical base, the inexperience and shortage of planners, the failure to base plans on an overall investment and development strategy, and the US embargo leading to a shortage of spare parts etc., produced a severe economic downturn. Consequently, 'In 1963 a reformulation of the development strategy of the Cuban economy began' (Brunner 1977: 35).

The Soviet model of central planning, exclusively concerned with production targets, ignoring the moral and ethical dimensions of socialism, was not seen as a model to emulate.

> [F]or both Che [Guevara] and Fidel [Castro], socialism was not simply a matter of developing a new way of distribution, it was a question of freeing people from alienation at the same time.
>
> (Valdéz Gutierrez 1996: 20, my translation)

Motivated to avoid the emergence of a class of professional politicians intent only on remaining in office, 'channels for the expression of the popular will' (see quote from Che Guevara above) were to be through the 'mass organizations': trades unions, neighbourhood-based Committees for the Defence of the Revolution, the Federation of Cuban Women, the Association of Small Farmers, the Federation of University Students, the Federation of Secondary School Students, the Communist Party, and various other groupings (Cole 1998: 27). For Castro, these organizations 'constitute the great school that develops the *consciousness* of the millions and millions of workers, men and women, old people, young people, and children' (Castro, quoted Medin 1990: 156, emphasis added).

The economic strategy intended to complement this approach to participation emphasized producing ever bigger *zafras*, the sugar harvests, to reinvest the surpluses to finance industrialization to escape sugar dependency. The apogee was intended to be a record 10 million ton *zafra* in 1970.

> If we fail to achieve the 10 million [tons] . . . we shall suffer a moral defeat . . . This time it is the whole nation who is to make a superhuman effort to reach a goal which shall be raised as the banner for what we stand for, the banner of socialism, and for which we have fought with a determination with which all revolutionaries need to fight to achieve their goals.
>
> (Castro 19 May 1970, quoted Brunner 1977: 108)

The target was not met; 'we alone are the ones who have lost the battle, the administrative apparatus and the leaders of the revolution are the ones who have lost the battle' (Castro 1972: 296, my translation). And Fidel Castro offered to resign.

The economic development strategy had to be rethought. Cuba joined the East European trading block, the CMEA, in 1972, and by 1975 adopted a system of central planning based on the Soviet model, the SDPE (Sistema de Dirección y Planificación de la Economía), which by 1980 applied to some 95 per cent of the economy (Hernández and Nikolenkov 1985).

The planning exigencies and bureaucratic tendencies of economic coordination with the CMEA/SDPE were to be reconciled with the ethics of the revolution through *Poder Popular* (popular power). Democracy was no longer to be left to informal mechanisms of the mass movements, but was to be institutionalized. Poder Popular, or more specifically the Organs of Popular Power (OPP), decentralized the management of productive and service enterprises to the areas and constituencies that they served: a hierarchy of political institutions, from the *circumscripciónes* (in Britain these would be electoral wards, the constituency that elected representatives), to the municipality, the province, and the national assembly. For a detailed analysis of these institutions see Cole 1998: 36–9.

The SDPE planning system relied upon enterprises' returning a profit within the parameters of planned prices, and consequently those activities that were relatively less profitable tended to be ignored or given lower priority. And within the CMEA Cuba remained essentially a sugar producer, a continued dependency that militated against a national development strategy of industrialization: consequently 'projects weren't being finished and [enterprises] . . . wanted to meet the plan in terms of value but not quality . . . this system was not only incapable of running the economy efficiently but of overcoming underdevelopment' (Castro, quoted Reed G. 1992: 256, 102). In addition there was the problem that with central planning there was very little scope for the Organs of Popular Power to allocate resources, and in addition the volunteer representatives often did not have the time, knowledge or skill to find their way through the bureaucratic labyrinth of the state.

It became apparent that a quantitative system of economic planning was incommensurable with a qualitative political participation and socialist development. And Cuba entered a period of 'rectification' in 1986: the 'Campaign of Rectification of Errors and Negative Tendencies'.

> The most serious error of economic policy in practice between 1975 and 1985 [the SDPE] was undoubtedly its reliance upon economic mechanisms to resolve all the problems faced by a new society, ignoring the role assigned to political factors in the construction of socialism.
>
> (Castro 1987: 13, my translation)

Central planning was ended for political reasons: 'little by little we began to recover the idea that the revolution was not only a matter of a more just distribution of wealth, but also a spiritual project to release people's *creativity* and give them a degree of participation in society' (Blanco and Benjamin 1994: 28, emphasis added). However, an alternative process of economic coordination was stymied by the collapse of the Soviet bloc, and in 1990 Cuba entered a 'Special Period in Time of Peace' (on this period see Cole 1998: 39–59).

Socialist development, Cuban style

Socialist development is a process of people's empowering themselves, a process facilitated by political activists, and hitherto the benign paternalism of Fidel Castro and the Cuban political leadership has been very important in this regard, in experimentation with democratic procedures. Cuba's electoral law stipulates that elected officials cannot be full-time politicians, a legacy from pre-revolutionary times, when 'corruption was the main course of political diet' (Reed G. 1992: 114).

Part of this process has been the trend towards equality.

> For there to be 'true' democracy the exploitation of man by man has to be ended. I am absolutely convinced that while there exists enormous inequalities between people is not possible to have democracy.
>
> (Castro, quoted Muñiz 1993: 1, my translation)

However, in the 'Special Period', because of shortages almost everyone turned to the black market to survive, and because the black market traded in illegal dollars, almost all Cubans were 'criminals'. Hence the dollar became legal currency. Henceforth, people with access to dollars – Cubans with relatives in the United States, taxi drivers, bar tenders, people working in the tourist industry or in export industries – are rich. Because of planned prices over the past thirty-nine years, the cost of living has not kept pace with world inflation. In mid-1998 the average monthly wage was about 240 pesos, and with a dollar–peso exchange rate of 22 pesos, a $10 tip to a waiter is almost a month's salary. And in a society that has always placed a premium on education and public service the resultant inequality has created tensions and disillusionment.

Continued progress in Cuba, the better fulfilment of individuals' changing social potentials – socialist development – rests on a deepening class-consciousness: a self-awareness of the individual's place in society, which is an *ideological* struggle – ideology that explains the experience of people being saved by health programmes, benefiting from educational provision, being gainfully employed, not being subject to the arbitrary repression, etc. '. . . the task of the revolutionary is first of all to arm people's minds, arm their minds! Not even physical weapons can avail them if their minds have been armed first'

(Castro, quoted Medin 1990: 5). There is not space to deal fully with the evolution of Cuban development, and the ongoing process of class-consciousness in Cuba, but see Cole (1998), especially Chapter 8: 'Cuban development – the future'.

> [S]ocialist democracy mandates the elimination of class divisions between the exploiter and the exploited . . . Liberty for one means oppression for the other . . . The struggle against [cultural and ideological] deformations of the past is a long-term project, not one to be accomplished overnight with the defeat of the capitalist state.
>
> (Evenson 1994: 24)

Part IV
Knowledge

Conceptions of knowledge

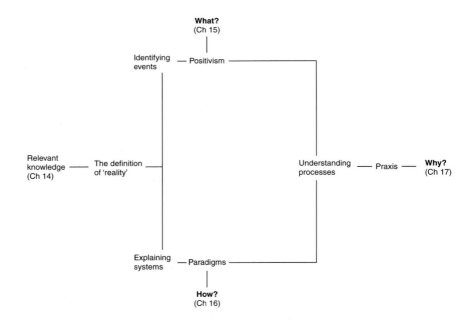

14 Knowledge

Explanation

> The phenomena we describe and purport to explain appear to be
> constructed according to hypotheses derived from our own fallible senses,
> cultural traditions, social expectations and limited technological powers.
>
> (Rose 1997: 69)

In the Prologue, written to underline for the reader the theme of this book
– what *can* we know about our existence – this quote was cited to highlight the contingent nature of 'truth'. To be able to say something about
the world we need to know the context of the question being asked. In
particular we need to know why should a 'newly intelligent [in evolutionary
terms] upright ape' try to make sense of an 'utterly incomprehensible . . .
universe' and try to encapsulate 'the sheer vastness of that universe in its
tiny brain case' (Stewart and Cohen 1997: 5). In short, *why* are we curious?

We have already seen in Parts I, II and III of this book the range of
possible, coherent explanations and understandings of the economy, the environment and development. And the theoretical perspectives through which
these aspects of experience are rationalized rest upon assumptions about the
nature of that experience: assumptions that imply beliefs about the dynamic,
the motivation, the causation, of social activity. But what structures these
beliefs? This is one of the 'most fascinating riddles of the 21st century: the
riddle of the human mind' (Wolfe 1997: 10).

> Everybody has a world view. It is always the most important determi-
> nant of what one sees, and it is almost impossible to describe.
>
> (Rose 1997: 2)

Facts do not speak for themselves, and understanding is inevitably dependent on non-scientific factors: intuition. And yet 'there must be a reality
which *at least in part* is independent of our experience of it and our asser-
tions (including our theories and models) about it' (Hodgson 1993: 12,
emphasis in original). It is not that ideology, the beliefs that structure our
understanding, is distinct from science; rather there are different *ideologies* of

science: 'visions are the basis on which theories are built' (Sowell 1987: 14) – visions that interpret that 'reality independent of our experience' differently. Rarely do theories change because of a change in a person's vision; rather, it is a change in the experience being reported. For instance, in Chapter 7 we saw how Meadows (*et al.*), working within an *egocentric* conception of the environment, emphasized the importance of population pressure on environmental resources, and when their apocalyptic predictions of starvation, shortages, overcrowding and pollution did not come to pass, the same vision was used to reinterpret experience to account for the unexpectedly more favourable conditions of social existence. And with *ecocentric* analyses, where the environmental (technical) parameters of social existence need expert management, although different schools within this perspective emphasize different aspects of the technical–environmental relationship, rarely is the ideology of ecocentrism addressed. The environmental context of social life is assumed, and 'although the environment *must* be considered to be socially constrained, there are aspects of the environment that are more . . . socially constructed than others' (Blaikie 1995: 9–10, emphasis added). It is left to 'experts' to define the social parameters of how (not *why*) we depend upon the environment.

Egocentric and ecocentric analysts are using different theoretical languages to address the *same* environment, the same 'reality independent of experience'. While often using 'the same vocabulary' each vision uses 'words differently': analysts working within different perspectives speak from 'incommensurable viewpoints' (Kuhn 1970: 200). Analysts have different perceptions of the *same* reality, effectively different realities, and the communication problems between 'members of different language communities [are] problems of translation' (Kuhn 1970: 175).

We illuminate reality through conceptual lenses, which provide meaning to our experience, and these lenses define what we see, affecting our intuitions and our activity. There is *no* neutral way of understanding life; 'no set of conceptual lenses allows us to see all reality' (Wells 1996: 2). But to us this distorted representation *appears* real: it *is* our experience.

> [R]eality is conceptualized by humans in various ways, resulting in many 'versions' of the world, none of which can be considered the single absolute version.
>
> (Hodgson 1993: 14)

However, we do check conceptions and theories against our experience of the world, but 'comparison of theory with experience is *not* comparison with unconceptualized reality . . . it is comparison of one or another version with the version we take to be "experience"' (Putnam 1983: 162). And in this sense, knowledge is power. Once a particular representation of social life becomes culturally, and therefore politically, dominant, social life is organized to reinforce this bias.

A definition is a meaning that is assigned to a situation and thus the mould through which all decisions are cast. The definition also partially defines what the individual will believe about reality. It is really a *value statement* about the nature of the problem and, by implication at least, a prescription for the solution.

(Wells 1996: 5, emphasis added)

However, the production of knowledge by intellectuals is legitimated as truth by ideologies of science, and although intellectual hegemony confers power on the ruling political elite; often the knowledge producers, in universities, schools, research institutions, the press and media, etc., are unaware of the bias they are unwittingly passing on as received knowledge. 'Science . . . is not "neutral". Its objectivity is only skin deep . . . shaped at least in part by our own social expectations and philosophy' (Rose 1997: 53).

Science disempowers: science has taken over from religion in justifying authority. Rationality is a guiding principle for political choice, and contrasts with the pre-Enlightenment trust in divine providence: 'what is taking place is an exercise of power through subtle but effective instruments for the formulation and accumulation of knowledge' (Leach and Mearns 1996b: 16). Scientific inquiry does not give facts but interpretations; not answers but an assessment of the feasibility of particular courses of action in terms of certain assumptions about life. Science studies the consequences of particular beliefs about the nature of existence.

By defining what is acceptable as evidence certain privileged methods . . . act to exclude . . . data . . . certain questions remain unasked, and certain types of evidence are ignored or dismissed as invalid.

(Leach and Mearns 1996b: 14)

Scientific experiments allow us to identify the effect of a particular course of activity, and theories place experience in a conceptual framework that allows us to understand, give meaning, and choose how to act, within the belief structures that define our 'real' world. Scientific, absolute truth 'has gone the way of the dodo' (Stewart and Cohen 1997: 36). Objective reality is *always* perceived subjectively, and the 'real' world beyond individual perception is too complicated for our minds. The actual world of individual experience is not the same thing as the infinitely complex real world of social existence.

What we believe the world to be, *ontology*, which defines the questions we wish to ask, determines how we study and understand existence, *epistemology*. Fundamental to explaining social activity and social change, is theorizing *why* people behave in particular ways: what determines human motivation? What is human nature? And 'the theoretical framework that surrounds our observations . . . underpins everything we [as analysts] do' (Rose 1997: 26). And in one way or another we are *all* analysts: why?

The evolution of self-consciousness

> . . . creatures look up into the night sky and see the stars. But we stare
> at them, wonder how many there are, wonder how far away they are,
> wonder how they got there, wonder what they are made of, wonder –
> indeed – why they are there at all . . . We see no evidence that any other
> creature looks outside its *personal* universe in this manner.
>
> (Stewart and Cohen 1997: 6, emphasis added)

Humans have developed a personal, *self*-consciousness, unequalled in the
animal kingdom. And this unparalleled self-perception and self-awareness
arose 'in the course of the ascent of life from the primaeval brine because
it confers spectacular survival advantage' (Denton 1993: xi–xii). Human
features have origins in earlier life forms, and there is a gradient from simple
perception up to human mental capacity. And with the increase in the size
of the brain has come man the toolmaker, producing tools for an *imagined*
eventuality, and the capacity to socially manipulate and control natural forces
changing the evolutionary path of the human species, and the emergence of
civilization.

> Isn't it strange that the animal we used to be developed into the crea-
> ture we are now.
>
> (Stewart and Cohen 1997: ix)

With an increasingly complex consciousness came self-awareness, people
being able to imagine what they will be doing: humans have a sense of
self. We perpetually react to 'our estimate of what the world will be rather
than what it is right now' (Stewart and Cohen 1997: 7). We are able to
have a vision of past experience, and replay it to imply a changed, preferred
outcome: we can *choose*. And herein lies the survival advantage: being able
to exercise options and not repeat past mistakes. We are not passive observers
of reality; we construct maps of that reality, allowing us to decide how to
interact with that reality. And we are able to share these lessons, and pass
on the wisdom of experience, through language. The symbolic representation of reality in language means that not everyone has to experience the
same thing to learn from that experience: we don't have to keep inventing
the wheel.

At some point the capacity for symbolic expression was genetically
imprinted, dictating the organization of the brain. And with this capacity,
henceforth, humans were able to change the conditions of existence faster
than genetic evolution. Through language people can process non-genetic,
non-instinctive, information, and consciously manipulate the environment
through which they survive: 'culture has changed the direction of genetic
evolution and greatly accelerated it' (Stewart and Cohen 1997: 138).

In fact nature [biology] and nurture [society] are not exclusive. Every attribute is influenced by both ... stone tools, agriculture and private property all had an effect on society and in turn on evolution.

(Jones 1996: 214; 1994: 301)

We are in control of, and can make choices about, our future only because we have *minds*, which we can make up or change. We have two unique attributes: 'knowing the past and planning the future (Jones 1994: 300). But the mind is not a 'thing': it is a 'process'. It has no location. An automobile moves, but where is movement located: in the wheels, in the engine, in the fuel? It is in all, and none of these places. Similarly, a thought has no location. Many animals have brains, but minds appear to be less common, and are of a higher order, facilitating sophisticated intelligence: 'the transition from brains to minds can be traced to the time when animals came up with non-genetic routes to protect their offspring' (Stewart and Cohen 1997: 27). The emergence of the 'symbolic representation of reality': language. And central to the evolution of the mind is the cultural context of teaching and learning. New forms of behaviour are invented by the mind, and although genetically influenced, the mind 'permits each combination of genes to have multiple expressions and offers alternative solutions to many problems, even within ... [an evolutionary] lifetime' (Denton 1993: 137).

Self-consciousness, the ability to imagine possible scenarios with yourself as the central actor, presenting several options allowing choice between alternative possibilities, is an effect of the genetic evolution of the brain and the cultural context of human social existence. There is not a dichotomy between the biological (nature) and the social (nurture): between the body and the mind. To believe that the mind is distinct from the body, is to allow for the possibility of a 'conscious afterlife and to many kinds of supernatural, paranormal and other worldly beliefs' (Sperry 1980, quoted Denton 1993: 11). The mind and the body are different aspects of a single, complex entity.

We humans are made from the same atoms as the rocks, water and air around us. The inevitable conclusion is that it is not the ingredients that differ but the way they are arranged.

(Stewart and Cohen 1997: 13)

At any instant the brain processes prodigious amounts of information from our senses: taste, smell, hearing, seeing, feeling, etc. But we are aware only of particular features of our surroundings; our minds filter out the important aspects for choice. Experience is recorded selectively.

Psychological experiments have shown that people's memories of specific events depend very largely on their *general assumptions*, on the conceptual structure that organizes their mind. Broadly, only items that can fit into the conceptual spaces within one's mind can be stored there;

items that do not fit are 'squeezed' into (or rather out of) shape until they do.

(Boden 1992: 243, emphasis added)

The choosing, creative mind transcends the physiological entity of the brain, and consciousness is emergent, it is a process, an indeterminate non-material feature of human life, so that potentials can be assessed: having a mind is having goals to achieve. What we know reflects our image of the world, an image contingent on our creative ambitions, in turn reflecting our potentials (or what we think our potentials might be).

Creativity is a paradox. How do we create: how, apparently, is something brought into being out of nothing? Is it insight, romantic inspiration, intuition? Is it something that only a chosen few have? – 'despite the elitist claims of inspirationalists and romantics alike – we all share some degree of creative power' (Boden 1992: 12). Being creative is not the same thing as being original in any absolute sense, doing something or thinking something that no one else has ever thought of or done before. It is about people fulfilling and achieving *their* potentials, potentials that change with social experience as people become aware of what they might achieve in concert with other people: 'creativity requires no specific power, but is an aspect of intelligence in general' (Boden 1992: 24).

Some people might have particular, innate, mental or physical abilities, or are able to develop particular conceptual structures – people described as 'gifted' – but these are more efficient variants of abilities which we all possess. To be creative requires knowledge. People must understand their activity, and then make the creative leap: 'Creativity is 90% perspiration and 10% inspiration'. As such, creativity requires commitment and interest, but also self-confidence. To pursue new, possibly original ideas, make mistakes and withstand criticism requires courage. Creative insights clearly rely on prior thought processes of noticing, remembering, intuitively correlating apparently unrelated phenomena, etc. – 'I have had my solutions for a long time, but I do not know yet how I am to arrive at them' (Carl Gauss, mathematician, quoted Boden 1992: 25) – but the actual creative insight appears to be unconscious. It is an emotional act: people intuitively following hunches, guided by beliefs.

The formalized language of scientific analyses of particular aspects of social existence are rendered contextually relevant to social experience as a whole through the medium of beliefs – beliefs that account for the dynamic, the causative mechanism, the determinant, of social experience. A person's belief structure provides a sense of order, and an understanding of experience, allowing people to make choices about how to fulfil their social potentials: to think ahead, to predict the consequences of their behaviour and anticipate what other people will/might do.

Mental events do make a difference to what physical events occur, which cannot be fully explained in terms of physical laws ... *the mind matters*

... mental events make such a difference by virtue of informal plausible reasoning, which cannot be completely formalized, and which requires consciousness.

(Hodgson 1993: 6, emphasis in original)

Rationality, 'informal *plausible* reasoning', is founded on beliefs, and is the guiding principle for choice. Culture, and the resultant belief structure that generally rationalizes social life, tends to justify the social status quo. Particular knowledges are privileged and institutionalized, and social conflict is reflected in competing theoretical perspectives, which rationalize social activity according to competing systems of beliefs, providing alternative, conflicting analyses of social existence: 'science . . . is *socially organized* hypothesis-making' (Rose 1997: 66, emphasis in original).

Science is a social activity reflecting visions of life, world views, modes of thinking, 'which establishes a general frame of reference for more specific hypotheses . . . within individual research fields' (Haila and Levins 1992: 105). The distinction is between general perspectives and particular schools of thought. The analysis of a particular problem, rationalized through a school of thought, only has meaning when contextualized to the underlying general perspective.

[S]cience is a human activity and therefore is a proper subject matter for social studies. Science is one of the things that people do, and therefore it can be studied in the same way that other human activities are studied.

(Diesing 1982: 1)

Necessarily, all scientific, theoretical work addresses phenomena that have been abstracted out of the context of their existence, and hence phenomena are intellectual constructs – concepts – that have been invented to help understand the world. But they are not arbitrary; they are derived from interpretations of actual, lived, experience. And it is to how experience *might* be rationally interpreted that we now turn.

Experience and belief

How might we coherently begin to socially rationalize our individual experience? Essentially people need an understanding of why people, including themselves, behave in particular ways. What determines human motivation? What is human nature?

Is it individuals' biological, genetic inheritance – nature – that predisposes them to certain attitudes and behaviour?

A society can be described only as a set of particular *organisms*, and even then it is difficult to extrapolate the joint activity of this ensemble . . . that is to predict social behaviour.

(Wilson 1980: 77, emphasis added)

Or perhaps it is their social experience – nurture – that has taught people how to adapt to survive?

> Man is innately programmed in *such a way that he needs a culture to complete him* . . . Man is like one of those versatile cake mixes that can be variously prepared to end up as different kinds of cake . . . just as a cake has to be baked, so a baby has to be exposed to a specific, already existing, culture.
>
> (Midgely 1978: 286, emphasis added)

Now I know what people mean when they refer to me as a 'fruitcake'!

Alternatively, are a person's biological inheritance and their social experience interrelated and complementary?

> The properties of individual human beings do not exist in isolation but arise as a *consequence of social life*, yet the nature of that social life is a *consequence of our being human*.
>
> (Rose *et al.* 1984: 11, emphasis added)

If humans are understood to be biologically determined, individuals are like negatives waiting to be developed by society. Nature determines the image and social experience; nurture is the fixing solution.

Where individuals are believed to adapt to the needs of society, people are *tabulae rasae*, a blank sheet upon which society writes its message. Nurture determines the story, nature is merely the medium: the paper or the canvas.

Where either nature or nurture is the determinant, individuals are essentially passive, being moulded by forces beyond their control: genetics or social structure. But where people's biological inheritance is understood to be revealed only *through* social activity – indeed, people only become aware of what they can do when they interact with other people, and their potentials change *with* social activity – then people *can* be active participants in their own development. There is a dialectic between nature and nurture which is more than interdependence. Neither can exist, nor be empirically identified independently, and yet behaviour cannot be understood without being aware of the contribution of each part. And the interaction of nature and nurture produces results qualitatively different from the constituent parts. Human nature is emergent: a process *between* nature and nurture.

Human nature and understanding

The structure of the belief systems that underlie our explanations of social experience, facilitating understanding and allowing us to choose how to conduct our life to fulfil our potentials, ultimately reflects what we think people, including ourselves, *are*. And as we have argued above, our conceptual, theoretical, parameters condition the sort of questions we are able to ask.

If individuals are understood to be biologically determined, then what they choose to do is entirely a matter of how individuals interpret *their* lives and experience, and what they think their interests are. Individuals are essentially independent of society.

The only questions that can be asked are about what has happened, is happening, or might happen. And predictions of the future are merely extrapolations of the past, based on the assumption that the evolution of biologically determined human nature is so slow that it is unlikely to change in our lifetime. Hence some economists believe (assume) individuals to be utility maximizers, the parameters of which are individuals' biological endowment of tastes and talents, determining their subjective preferences.

But if individuals are understood to be socialized by their experience, by which they learn to adapt to the needs of society if they are to survive, then individuals are not independent of society, but dependent on society. The relevant questions are now not what occurs, but *how*. We are not interested essentially in what individuals choose to do, but how the social system presents certain alternatives, between which all people have to choose.

When individuals are believed to be independent choosers, scientific analyses are necessarily *reductionist*. The whole (society) is understood through the activity of the parts (individuals).

When individuals are believed to be dependent choosers, scientific analyses are *holistic*. The whole (society) conditions the activity of the parts (individuals).

As ontologies and epistemologies, both reductionism and holism are equally valid, scientific practice being legitimated by analogous beliefs in human nature: biological determinism or social adaptation. But in both of these approaches to human understanding, individuals are essentially passive victims of forces beyond their control: genetic inheritance, or cultural tradition.

However, where people are understood to be an emergent product of their biological characteristics *and* their social experience, a consequence of the dialectical interaction of nature and nurture, then people are neither independent of, nor dependent on society, but interdependent within society. People change through social experience. And if we can understand the process by which the emergent nature of interdependent people occurs, then people collectively can begin to actively participate in guiding their own future towards fulfilling their human potentials.

People need not be passive victims of forces that they do not understand: therein lies the road to bigotry and prejudice. Such a passive acceptance, a fatalism, about their future expresses itself in a religious faith in the benign intentions of supernatural forces, and social rituals emerge designed to appease the wrath of God or some other imagined supernatural deity.

Where people are considered as interdependent beings within society, then analyses are not principally concerned with identifying what, or explaining how phenomena come to pass, but *why*. If we are active, interdependent people, concerned to fulfil our social potentials, why have individuals' activity and behaviour produced a social context for existence that denies some people this opportunity? Their social position limits their ability to be themselves – to be creative – and denies their humanity. And such a denial either results in people internalizing their frustrations, possibly being expressed through psychological disorders, or drug abuse, or violence; or people may be conscious of others who share their predicament, and mobilize and organize to change society, allowing them to live a more human life – a life that allows them to fulfil their creative potentials.

> We cannot evade responsibility by pretending that our choices are dictated to us from outside or assume that doing nothing is acting wisely.
>
> (Haila and Levins 1992: 13)

Either way people are not able to meet their full potential, and society is the poorer.

> What matters is not whether we modify nature or not, but how, and for what purpose . . .
>
> (Haila and Levins 1992: 11)

The next three chapters go on to examine the science of asking 'what?', 'how?' and 'why?', looking at the process (why?) by which events (what?) are systematically (how?) related.

> [T]hings [events] and systems are perpetually constituted and reconstituted . . . out of processes.
>
> (Harvey 1996: 51)

15 What?

Identifying events: positivism

Events

All phenomena, sensual perceptions, exist as *events*: an occurrence, a thing, an abstraction, a feeling, etc. And all scientific investigation has to establish the facts of what happens. But these facts exist only as *conceptions*, which more or less help us to understand the world and thereby choose how to better organize our affairs. Atoms do not exist; the atom is a conceptual construction that enables us to understand the existence and movement of bodies (Haila and Levins 1992: 104; Silver 1998: 363–8). Similarly, a reductionist approach to social understanding, whereby individuals are considered to be independent beings – individual human atoms in the body politic – is a *mode of thinking*. The idea that 'society can be described only as a set of particular *organisms*' (Wilson 1980, quoted Chapter 14 above) is a particular vision or world view. The concern is not with *how* they got to be where they are or how they behave, or *why* they exist, but *what* they are and what they did. Facts are a selective representation of experience. We hold the distorted, conceptual mirror of theory up to the world, and record the image as 'reality'.

Reducing systems of how phenomena occur, and processes of why they exist, to a reductionist methodology of what the events are, what the components are, has been extremely powerful over the last three centuries, and 'has given us unrivalled insights into the mechanics of the universe ... And our experiments are productive ... Within limits our experiments are successful' (Rose 1997: 78). As a scientific approach it allows us to isolate the phenomena in question and alter potential variables one at a time, avoiding the difficulties of disentangling several coterminous variables and isolating particular causal dynamics that act in concert.

A problem is reduced to its smallest terms, with the fewest possible variables, an approach that is very useful for bounded, mechanical systems, where each part has an unambiguous role to play in the operation of the system. In such a conception of knowledge and reality two variables cannot be joined by more than a single link in each direction, and variables outside a particular discipline are deemed to be constraints or externalities, independent of the assumed variable.

> Thus, economic models of agricultural production link prices and yields but ignore the effect of these on erosion or pest problems that eventually alter the yield.
>
> (Haila and Levins 1992: 49)

Such an approach to knowledge allows for metaphysical, mystical or religious explanations to account for variables outside the scientific parameters of the analysis. Beliefs, potentially, are rendered commensurable with scientific analyses.

Where there are fixed relations between the parts, for instance the components within an internal combustion engine – each is designed and manufactured separately, and the working of the engine can be predictably anticipated based on known physical and chemical regularities – then even where the mechanism is very complex, reductionism is a valid methodology.

But such an approach can only be applied to social analyses, and therefore economic, environmental and development studies, if it is assumed (believed) that individual behaviour follows an unchanging, inevitable pattern: the 'parts', individuals, are all the same and are predictable. Human nature is believed to be innate, biologically determined. People are socially independent and complete, and compete within society to socially fulfil their biological potentials, and individuals' genetic endowment determines their superiority justifying inequality. People should justly receive what is (assumed to be) a product of their efforts. Individuals are believed to be hedonistic: '*incentives matter to all human behaviour*' (Anderson and Leal 1991: 10, emphasis in original). And because humans are independent beings, social phenomena are discrete events motivated by individuals' preferences, and hence 'knowledge cannot be gathered into a single mind or group of minds [e.g. planners] that can capably manage' (Anderson and Leal 1991: 4) social existence. To maximize our individual potentials we need to live in a free society.

Assuming that individual variables can be meaningfully isolated is asking 'what if' questions to illuminate experience. What would happen if the economy were perfectly competitive? What would happen to the environment if there were fewer people? What would be the outcome of social change if people were able to act in their own interests? But inevitably this becomes the way experience is perceived and conceptualized, and 'what if' questions become 'there are' answers.

> Effective experiments demand the artificial controls imposed by the reductive methodology of the experimenter, but we must never forget that as a consequence they provide at best only a very simplified model, perhaps even a false one, of what happens in the blooming, buzzing, interactive confusion of life at large . . .
>
> (Rose 1997: 28)

Reductionism reduces wholes to their parts, identifying the components that interact, defining essentially independent events that are correlated to imply a causative mechanism. The method is to describe the internal workings of the whole, correlate changes in the separate components, and derive hypotheses as to the relationship between discrete events. Note: there is a *relationship*, not *causation*, between variables.

Hence, as we saw in the section on the economy, for the subjective preference theory of value in economics, if people behave as assumed (believed), then price will fall as supply increases. The prediction is apparently concrete, testable, and self-contained. And the hypothesis is not meant to describe how this relationship takes place, what is the causation, the dynamic of behaviour.

> [T]heory is to be judged by its predictive power for the class of phenomena which it is intended to 'explain'. Only factual evidence can show whether it is 'right' or 'wrong' or, better, tentatively 'accepted' as valid or 'rejected' ... Truly important and significant hypotheses will be found to have 'assumptions' that are wildly inaccurate descriptive representations of reality, and, in general, the more significant the theory the more *unrealistic the assumptions* ...
>
> (Friedman 1953: 8, 14, emphasis added)

'Significant' theories abstract out all the important phenomena from the complex circumstances of daily life, and prediction is based on these alone: hence the theory is an 'inaccurate representation of reality'. And only empirical evidence is admissible; abstract reality, concepts, thoughts and feelings are irrelevant to scientific inquiry.

Objectivity and truth

'Objectivity' is defined by anyone else making similar observations seeing the same thing. But of course there is no way of actually knowing that, in future, similar observations will always be made: remember, phenomena are treated as independent events, relationships are established by correlating changes, and such changes are not a consequence of a theorized causation. The relevant data are quantitative, and are manipulated by statistical techniques to establish relationships between variables, so that: managers can compare revenue and cost data and establish the average return on investment; health planners can establish the relationship, say, between smoking and/or obesity and a range of diseases; economists might describe the relationship between the demand for a commodity and income, family size, age or ethnic background of consumers; educators might establish a relationship between ethnic background, sex, or income, and intelligence; etc. In each of these cases data will be collected, classified, and interpreted through techniqu
frequency distributions describing central tendency, dispersion, a

deviation, intended to establish the probability distribution and the confidence with which a relationship can be established. And because no causative mechanism is assumed – events are independent – forecasting the future can only be an extrapolation of past trends: and any change is a surprise.

Truth is always contingent; something is true until the predicted events fail to materialize: Karl Popper's *positivist* theory of science: '*for strictly logical reasons, it is impossible for us to predict the future course of history*' (Popper 1960: v, emphasis in original). Behaviour reflects knowledge and understanding, and Popper argues that because we cannot predict what we will know in the future, we cannot predict the future course of history.

But it is the history of *events*.

> [A] causal explanation of a certain *specific event* means deducing a statement describing this event from two kinds of premises: from some *universal laws*, and from some singular specific statements which we may call the *specific initial conditions*.
>
> (Popper 1960: 122, emphasis in original)

That is, in the case of a causal analysis, where causation is assumed rather than the correlation of events defining a relationship, truth has to be *induced* from axioms, not *deduced* from facts. That is, our observations are made with reference to an assumed notion of causation, and used to 'prove' these assumptions: the emphasis is on *verification*. But for Popper the statement of a universal law, or the assumption of specific initial conditions is not a known fact. Popper accepts that knowledge cannot just be the collection of facts, since facts are collected according to a conceptual interpretation of experience.

> Before we can collect our data, our interest in *data of a certain kind* must be aroused: the *problem* always comes first. The problem in its turn may be suggested by practical needs, or by scientific or pre-scientific beliefs which, for some reason or other, appear to be in need of revision.
>
> (Popper 1960: 121, emphasis in original)

The conceptualization of phenomena relative to a problem does not render scientific knowledge as truth. Rather, the problem should be stated as an hypothesis, which can be tested, and the hypothesis, if confirmed or at least not refuted, counts as 'truth' until it is disproved: the emphasis is on *falsification*.

Causation is irrelevant; the correlation of events is all.

Positivist social science then, is concerned only with testing hypotheses by choosing data that might show the predictions of a theory to be false.

Reductionism

This reductive, mechanical view of a world of components emerged as part of a process of social evolution between the sixteenth and eighteenth centuries. Previously the dominant vision or world view was *organic* (not mechanic). The culture of feudal life legitimated a social order where status was ascribed, with power based on the ownership of land, and people interacted through hierarchical relationships, their social position being defined at birth. The experience of social existence, the reality through which creative thought was manifest, was of a system where each person had a preordained place, and intellectual concerns were concerned to understand this natural, organic order – a form of thought that rationalized social existence and legitimized the social status quo.

> Knowledge of society is never a passive dogma; it is always active either in preserving or in destroying a social system.
>
> (Bernal 1969: 1025)

But societies change, and as labour services began to be remunerated in money in England from the twelfth century, and agricultural production was increasingly for the market, and manufacturing and towns emerged, power and social order was ever less a consequence of land ownership (Morton 1965, Chapter IV). And the social order could no longer be justified by appeal to an unchanging, natural order. Importantly, wealth and privilege were ever more a consequence of investing in production using an evolving technology, and market exchange. The notion of an 'organic, living, and spiritual universe was replaced by that of the world as a machine' (Capra 1982: 38).

Knowledge began to be legitimized by the thought of such people as Francis Bacon, who was concerned not with rationalizing a natural order, but with 'hounding' nature 'in her wanderings'; with making nature a 'slave' to be 'bound into service'; with 'torturing nature's secrets from her' – and to this end Bacon defined a scientific methodology for the discovery of this new knowledge (Merchant 1980: 169). There was René Descartes, who did not accept any traditional thought, and set out to replace it with a new conceptual scheme more relevant to the changing times – the 'foundations of a marvellous science' that would unify all knowledge (Vrooman 1970: 54–60).

> Acknowledging the crucial role of science in bringing about these far-reaching changes, historians have called the sixteenth and seventeenth centuries the Age of the Scientific Revolution.'
>
> (Capra 1982: 38)

The production of new knowledge by such people as Nicolas Copernicus, Johannes Kepler and Galileo Galilei reflected the decline of feudal authority and cultural hegemony, consequent on social change, which allowed such questions as 'Is the earth the centre of the universe?' (Copernicus), 'What are the laws of planetary motion?' (Kepler), and 'Can the laws of nature be expressed in mathematical symbols?' (Galileo) to be asked *for the first time* – a questioning of the 'organic' order that provoked the wrath of ecclesiastical authority.

Scientists were obsessed with measurement and quantification: defining *what* phenomena were. Rather than understanding the natural order and living in harmony with it, science and the resultant technology were intent on domination and control.

The changing significance of knowledge and changes in the social order, reflecting the tensions and contradictions in a social status quo where status was ascribed, went hand in hand with the emergence of capitalist organization of production. And the new social order had different concerns, which centred round the need of entrepreneurs to organize production and control the natural environment so as to maximize profit. Which fed back into the production of knowledge.

> ... classical physics was rooted in the economic and technological development of the 17th century ... [focusing on] such factors as communications, water transport, mining, armaments and ballistics ...
>
> (Gasper 1998: 150)

The practical needs of producers for a reductionist analysis of the material world spawned a *philosophical reductionism* (Rose 1997: 82–92) – that higher-order properties (wholes) are the result of, and secondary to, lower-order ones (parts). When this is applied to social studies, the implication is that the social order in general, and social activity in particular, is a consequence of independent individuals, who choose so to act (because they are socially independent beings).

As productive techniques became more advanced, the profitability of competitive enterprises rested on a more sophisticated analysis of the properties of the natural environment: 'modern industry ... makes science a productive force distinct from labour and presses it into the service of capital' (Marx 1974: 361). Knowledge reflects the conditions of social existence.

> By the mid-19th century it was already becoming evident that purely mechanical models ... were inadequate ... and much 20th century work in physics and biology has led to the questioning of reductionist assumptions ...
>
> (Gasper 1998: 144)

Such new phenomena as electromagnetism and radioactivity were outside the intellectual parameters of reductionist analyses.

In the physical sciences Einstein's theory of relativity argued that the physical properties of phenomena, such as space, time and mass, were not independent characteristics, and that Newton's laws of motion were a special case, an approximation, of the more general theory of relativity. Thus 'the basic concepts of classical physics ... needed to be modified, and apparently distinct features of the world are in fact deeply interrelated' (Gasper 1998: 154). And the rise of quantum mechanics in the 1920s, which denied that systems could be reduced to their parts, interrelations between the components reflecting the 'unbroken wholeness' (Bohm and Hiley 1975: 96, quoted Gasper 1998: 156) of physical systems, has also influenced social analyses (Zohar and Marshall 1994).

But just as the physics of Isaac Newton is strictly a special case of the more general theory of relativity, and is still the basis for the day-to-day understanding of experience, although the analysis of discrete social events cannot explain such phenomena as inflation, democracy, intelligence and social movements, 'the positive posture ... continues to this day to guide the efforts of practitioners of inquiry, particularly in the social or human sciences' (Lincoln and Guba 1985: 15). Such a conception of social reality is the 'dominant ideology' (Giroux 1981: 42) explaining and justifying the non-material aspects of social existence: social power. Indeed, the concept of falsifiability is so ambiguous and equivocal that as a scientific methodology 'its flaws both from the perspective of knowledge acquisition and from the perspective of error avoidance are dramatic' (Hausman 1988: 82).

Still, the scientific justification of experience as a series of discrete events can be justified by appeal to a belief in people as independent beings with a common biologically determined human nature: the calculus of hedonistic expediency. And even though we live in a world of growing social, political and economic interdependence, where even the 'most private behaviour of individuals affects and is affected by large scale social patterns in society as a whole ... [and m]echanistic notions that society consists of isolated units each blindly pursuing its own self-interest cannot cope with this interlinkage' (Zohar and Marshall 1994: 7, 8), observers and theorists continue to understand human behaviour as a product of independent, individual motivation.

It is 'common sense'. Our *actual* experience is of us apparently choosing how to behave, to exercise options according to our subjective preferences. And consequently our failures either reflect forces beyond our control, which inhibit freedom to choose; or the fault is ours. Ultimately our predicament is our fault.

Or is it?

> Common sense is ... inadequate ... All everyday experience is partial, and reflects the social location of the person both in the range of knowledge it confers and the perspective it gives.'
>
> (Loney *et al.* 1993: 341)

We now address analysts who, rather than ask 'what?', try to find out 'how?'.

16 How?

Explaining systems: paradigms

Systems

While all sensual perceptions, phenomena, appear as discrete events, it is the intellectual project of positivist scientists to correlate coincidental events to establish a tentative conception of reality. Events as empirical occurrences, events that can be corroborated by other observers (the definition of objective knowledge), and which have a material form, are the only admissible evidence. The abstract world of feelings, motivations, understandings and relationships is irrelevant to scientific analyses. Such events are correlated to establish a relationship in the attempt to disprove – falsify – hypotheses. And remember the plausibility of hypotheses, the accuracy with which reality is addressed, the extent to which theories mirror experience, is *not* an issue: 'in general, the more significant the theory the more unrealistic the assumptions' (Friedman 1953, quoted above).

Such a reductionist scientific agenda reflects the assumption, the belief, that phenomena in general, and individuals' social behaviour in particular, are independent events; there is no *necessary* causal link. But for other analysts it is assumed that there is a logic to how events are correlated: it is not serendipity, accident, or chance. Indeed, it is believed that the parts *only* have significance because of their position in the overall scheme of things, their place in the whole. Remember Piers Blaikie asserting that 'the environment must be considered to be socially constrained' (Blaikie 1995, quoted above)? The environment *cannot* be considered independently: a mechanistic vision of the universe is at best misleading, and at worst wrong.

> . . . to understand any piece of machinery you need to know not merely its composition but its role in the larger *system* of which it is a part.
> (Rose 1997: 95, emphasis added)

Such a perspective is holistic: what Ian Stewart and Jack Cohen call *contextualism*: 'Reductionism looks at the 'insides' of things, contextualism looks at the 'outsides' – in a *conceptual*, not a literal sense' (Stewart and Cohen 1997: 34, emphasis added). The intellectual parameters of science are not now restricted to a material reality; concepts are also real. By conceptualizing the relationship between events, by explaining the systemic relationship between phenomena, forecasting – predicting the future – can be more sophisticated than a mere extrapolation of past trends. We can anticipate the effects of present behaviour: if we continue to organize society so as to ever expand consumption and burn fossil fuels at an ever greater rate, global warming will change climates, affecting livelihoods, and rising sea levels will flood low-lying land areas, affecting lifestyles; if governments fail to manage economic distribution, and at the same time maintain full employment (and hence the money supply increases in line with the growth in production), then inflation and economic crisis will result; and so on.

Answers to problems are contingent on the behaviour of other parts in the system. Falsifying allegedly independent facts does *not* furnish useful knowledge. Facts cannot be true or false *in their own right*.

Paradigms

Historian of scientific thought Thomas Kuhn (1970) noted that scientists are not usually engaged in making, testing and falsifying hypotheses. Rather they are normally engaged in solving puzzles set by earlier researchers working in their field of enquiry.

> No part of the aim of *normal* science is to call forth new sorts of phenomena [falsification] . . . Nor do scientists *normally* aim to invent new theories . . . Instead *normal*-scientific research is directed to the articulation of those phenomena and theories that the paradigm already supplies.
>
> (Kuhn 1970: 24, emphasis added)

What is this concern with 'normal' science?

The intellectual parameters of scientific inquiry are not now the correlation of material phenomena, but the *paradigm* that is accepted by the scientific community. The concept of a paradigm implies: 'the entire constellation of beliefs, values, techniques, and so on shared by members of a given [scientific] community'; and 'the concrete puzzle-solutions which, employed as models of examples, can replace explicit rules as a basis for the solution of the remaining puzzles of normal science' (Kuhn 1970: 175). That is, researchers share a vision of the nature of the universe, and models of how that universe functions to 'solve the remaining puzzles of normal science'. It is a problem-solving approach to knowledge, *within* a shared perception, or vision, as to what those problems are. And scientifically trained 'experts' define what those problems are. As we shall see in Chapter 17 this is a very elitist theory of knowledge.

People face problems in their daily existence, and, applied to the social sciences, Kuhn's paradigms are a way of conceptualizing existence, conceptualizing the social system, in such a way as to solve such problems: solutions that generate scientific knowledge as a set of theories and rules – a paradigm. Paradigms are both a perspective *on* the world, and a model *of* the world. And because the paradigm successfully solves problems, it is generally accepted in the scientific community as *the* way to conceive of experience, and reality. Hence, even though different analysts may see or be aware of different stimuli owing to their different vantage points as researchers, those different stimuli when interpreted in terms of the same paradigm can produce similar sensations: people experience the same reality (conceptually defined).

> Notice . . . that two groups, the members of which have systematically different sensations on receipt of the same stimuli, do *in some sense* live in different worlds. We posit the existence of stimuli to explain our perceptions of the world, and we posit their immutability to avoid both individual and social solipsism.
>
> (Kuhn 1970: 193, emphasis added)

People do not experience independent realities (solipsism), because they have common parameters to their perceptions: they share a common paradigm. And the body of scientific thought, the extant paradigm, the dominant vision, is learnt by new practitioners in educational institutions: it is expert technical knowledge. A field of knowledge becomes more scientific, the more researchers share the same vision. '[I]n the early stages of the development of any science, different men confronting the same range of phenomena . . . describe and interpret them in different ways' (Kuhn 1970: 17), but over time different explanations are more or less useful, and a paradigm emerges: but 'a paradigm . . . need not, and in fact never does, explain all the facts with which it is confronted' (Kuhn 1970: 17–18). So it would not pass the test of falsifiability. It is a question of the plausibility of the explanation for *how* phenomena occur, not a question of describing or predicting *what* happens.

> Paradigms gain their status because they are more successful than their competitors in solving a few problems that the group of practitioners has come to recognize as acute.
>
> (Kuhn 1970: 23)

Revolutionary science

Normal science is the process of extending the explanatory scope of paradigms to better understand the systemic relation to which they apply, always bearing in mind that 'the paradigm sets the problem to be solved' (Kuhn 1970: 27). Problems are a reflection of the intellectual parameters of the paradigm. However, when new or unsuspected phenomena are revealed by research, or experience poses new problems, the paradigm (as a model) has to be revised to account for these anomalies. Subsequently, researchers are able to account for a wider range of phenomena, and explain systemic relations between phenomena more completely. For instance, with the need to address environmental issues and problems that do not neatly fit into the traditional disciplinary structure of scientific knowledge, such as acid rain, a variety of paradigms are brought into play, including 'engineering, chemistry, meteorology, biology, medicine, agriculture and mathematics' (Irwin 1995: 48). But when the anomalies multiply with research experience, and paradigms are unable to explain problems and solve puzzles, then the paradigm (as a world view) is in crisis, heralding a period of revolutionary science.

> Though there is still a paradigm, few practitioners prove to be entirely agreed about what it is. Even formerly standard solutions of solved problems are called into question.
>
> (Kuhn 1970: 83)

Research work concentrates on developing a new world view – the period of *revolutionary science* – to account for and explain these anomalies. Normal research is cumulative within a vision of reality: revolutionary (or extraordinary) research changes perceptions of that reality, and 'after a revolution scientists are responding to a different world' (Kuhn 1970: 111). The perception of reality reflects both the experience of that reality, and what people expect – believe – that reality to be. The culture of beliefs in society takes a long time to change, and the practice of research professionals is similarly recalcitrant, as textbooks have to be rewritten, the curricula of training courses have to be revised to reflect changing reality, and the priorities, methods and expectations of practising researchers have to be reviewed.

> Facts then are collected to be placed in the kaleidoscope of theory, and our perceptions of them are constantly transformed by the shaking of that kaleidoscope.
>
> (Rose 1984: 4)

However, and importantly for the concerns that will be raised in Chapter 17, the issue is essentially a logical one, to be solved within an intellectual discipline by experts; and eventually the 'entire profession or relevant professional subgroup' (Kuhn 1970: 144) will be converted to the new way of seeing science and the world. And the dynamic of change is human adaptation to the newly conceived problems of social existence.

The path of scientific development and the emergence of new knowledge is determined by pragmatic expediency. And when the scientific community fails to solve the puzzles that the extant paradigm defines as problems, then in the subsequent period of scientific revolution, not only is the way in which, *how*, puzzles might be solved revisited, but also the belief that these really *are* the important puzzles/problems is questioned. The new reality highlights revised priorities.

In Chapter 12, in discussing the reconsideration of structuralist conceptions of development in the light of more than twenty years' development policy intended to improve people's livelihoods in the Third World, after which on the whole the poor have got poorer and there are more of them, the concept of development is being fundamentally rethought. And to this end there is a theoretical impasse. There is a scientific revolution in progress: the chrysalis of a new paradigm is about to burst open.

Postmodernism

At this stage in the emergence of new theory perhaps the most vociferous controversy is over defining what the relevant problems *are*. Just as paradigms are a response to pragmatic expediency, so is objective truth. Thomas Kuhn noted above that with a change in paradigms comes a change in perception, the same stimuli generate different sensations, and analysts have to 'respond to a different world'. Postmodern theorists take this assessment to its logical conclusion, and insist that we have to abandon *any* attempt at a true understanding of the world. Knowledge is relative: it is only significant in the context to which it is a rational response.

> [T]he social scientist is . . . bound to his or her social and historical identity . . . Social knowledge is accordingly always partial . . . and a response to historically specific social conflicts . . . The social sciences have been deeply implicated in struggles over sexuality, gender, race, ethnicity, work, politics and education . . .
>
> (Seidman 1998: 348–9)

Postmodernists are confused paradigm theorists.

Life has changed so much that the mental constructs that hitherto have given meaning to experience have become anachronistic. And there are now postmodern tendencies in film, theatre, dance, music, education, broadcasting, language, art, architecture, philosophy, theology, psychoanalysis, historiography, science, cybernetic technologies, cultural lifestyles, etc. (Hassan 1987: xi; Hoggart 1998, Chapter 15), as well as in development theory.

Such a sea change in perception suggests a major turning point in social existence, the harbinger of fundamental changes in social organization and a challenge to traditional forms of authority, similar to the Enlightenment and the Scientific Revolution between the sixteenth and eighteenth centuries. 'As to the sense of Authority . . . where is the watershed? Perhaps at the end of the last war, there or thereabouts . . . The Authorities . . . have almost all gone. The contemporary ground-level is *relativism*' (Hoggart 1998: 4, 6).

It is difficult to relate thought to experience, knowledge is power, and 'the role of the intellectual . . . [should] not be that of establishing laws or proposing solutions or prophesying, since by doing that . . . [the intellectual] can only contribute to the functioning of a determinate situation of power' (Foucault 1991: 119). And for relativists, with no theory of social power, it is impossible to make choices between competing interests. Indeed, for Baudrillard, there is *no* reality beyond individual experience, and since social experience is gleaned from images in the mass media and literature, the image is reality, which is 'sheer nonsense . . . one of the silliest ideas . . . reality is purely a discourse phenomenon' (Norris 1992: 15, 11, 16).

The post-war era has seen huge social transformations.

The biggest mass strike the world has ever known occurred in France in May 1968. The student protest, beginning with a few hundred students, on 10 May, protesting over relatively minor issues such as poor conditions and authoritarian rules on university campuses, erupted into a mass strike of 10 million workers following the brutal suppression of student unrest by the riot police. On 13 May the trades unions called a demonstration, and over 1 million people took to the streets in Paris. The following day workers occupied the Sud Aviation factory over pay and conditions; on 15 May Renault workers near Rouen occupied their workplace, and within a week millions of workers were on strike or occupying their workplaces. Aerospace, car factories and all branches of industry, shipyards, railways, buses, banks, postal services, coalmines, museums, theatres, football grounds, radio and TV stations, were all affected or closed (Birchall 1987). The 'revolutionary rehearsal' (Barker 1987) of 1968 in France found echoes around the world. The year 1968

> was a year that marked an entire generation on every continent. Long before 'globalization' became a buzz-word in the culture of free-market politics, the events of 1968 had globalized political radicalism as part of a struggle to change the human condition for ever.
>
> (Ali and Watkins 1998: 7)

'. . . the worldwide revolution of 1968 (which in fact went on to 1970)' (Wallerstein 1996a: 209) saw independent struggles connected by a common concern with the legitimacy of political authority erupt in the United States, the Soviet Union, Czechoslovakia, Japan, Greece, West Germany, Italy, Spain, Poland, Brazil, Mexico, the United Kingdom, Yugoslavia, Israel, China, and Pakistan (Ali and Watkins 1998).

And the tide of people resisting the political status quo continued: in Chile in 1972–3 (Gonzalez 1987), in Portugal in 1974–5 (Robinson 1987), in Iran in 1979 (Poya 1987), in China, Germany, Poland, Romania, South Africa, Russia in 1989–90 (Simpson 1990), and in the rest of the Communist Bloc of Eastern Europe in 1990–91 (Callinicos 1991; Echikson 1990). On this whole period see Harman (1988). And at a more local level, almost invariably, internationally electorates' participation in parliamentary and local elections is declining. People are increasingly not prepared to be treated as political, independent individual consumers, who are enfranchised in choosing between political parties, in the same way as a shopper chooses between different brands of soap powder. As a consequence people's political participation has increasingly taken the form of single-issue social movements: race, gender, environmentalism, health, sexuality, disability, education, ethnicity, landmines, fair trade, poverty, transport, etc.

> [T]he world revolution of 1968, completed in 1989, involved a process of irreversible shift in collective social psychology. It marked the end of the dream of modernity ... the end of the faith that the state within the capitalist world-economy could serve as the facilitator and guarantor of steady progress ...
>
> (Wallerstein 1996b: 236)

But often the trend has not been towards participation in protest, but towards a reliance on radical expertise. And such a political reaction reinforces the importance of intellectuals, institutions, and management.

> ... knowledge is special and concentrated ... the best conduct of social activities depends on the special knowledge of the few being used to guide the actions of the many ... along with this has often gone a vision of intellectuals as disinterested advisers.
>
> (Sowell 1987: 46)

Intellectuals can be disinterested because the problems are perceived to be logical ones, to be solved by an appropriate management strategy based upon technical expertise.

However, it seems that the political institutions that bolster the social status quo are not commonly driven by disinterested logical persuasion. With no alternative political avenue for dissent, the collapse of old certainties, and the emergence of a culture that can only apparently promise insecurity, conflict, intolerance, a moral vacuum, and competition, people look to their own interests, adopting a 'pragmatic approach to knowledge that assumes ... no neutral agency to resolve differences and that we must struggle ... to negotiate identities, norms and common understandings' (Seidman 1998: 296). Concomitantly, intellectuals working within a theory of knowledge and a definition of objectivity based on pragmatic expediency, postmodernists and confused paradigm theorists cannot compare and contrast different relative truths. The views of fascists and anti-fascists, white supremacists and anti-racists, male chauvinists and feminists, etc. are all equally valid.

And such a moral confusion, particularly by 'the new middle class in retreat from the values of the 1960s' (Rees 1998: 297; Callinicos 1989) in intellectual debate, produces analytic works that, to try and rationalize a conceptually inchoate discourse, are 'obscure, excessively convoluted, pseudo-scientific claptrap. People may think that postmodernist ideas are "difficult because they are deep", but if they seem incomprehensible, it is for the very good reason that they have nothing to say' (Harman 1997) – what Gross and

Levitt categorize as 'unalloyed twaddle' (Gross and Levitt 1998: 43). Alan Sokal and Jean Bricmont, in their book *Intellectual Impostures* (Sokal and Bricmont 1998), address the

> fascination with obscure discourses; an epistemic relativism linked to a generalized skepticism toward modern science; an excessive interest in subjective beliefs independently of their truth or falsity; and an emphasis on discourse and language as opposed to the facts to which those discourses refer (or, worse, the rejection of the very idea that facts exist or that one may refer to them).
>
> (Sokal and Bricmont 1998: 173–4)

Alan Sokal describes himself as a 'stodgy old scientist . . . who believes naively that there exists an external world, that there exist objective truths about the world, and that it is my job to discover them' (Sokal, quoted Rees 1998: 22).

In Chapter 17 we turn our attention to defining the 'real world' of social experience, and to asking 'why?'.

17 Why?

Understanding processes: praxis

Processes

All phenomena exist as events, and all are related to other events, which positivist theorists try to deduce from correlation, believing events to be *independent*, causation being unknowable; and paradigm theorists attempt to plausibly induce relationships from a theory of causation, believing events to be *dependent* on each other. For the former experience is *described*, and objective knowledge reflects shared perceptions of reality; for the latter revealing the *systemic* relation between phenomena is the proper concern of trained intellectuals, and what is objective reflects the plausible consensus between experts as to what is reality.

However, the systemic relation between events changes, sometimes quickly, sometimes more slowly, changes which for paradigm theorists are surprises, to be explained by a period of revolutionary science. Alternatively another question could be asked: What are the particular conditions that led to the evolution and change of the systemic relation between phenomena? Ultimately, in the social sciences, *why* do people behave as they do? What is the *process* of human life?

> Like all animals . . . we humans have interacted with other members of our species and changed ourselves: our sense of identity has changed recursively as each interaction has built on the results of the previous ones.
> (Stewart and Cohen 1997: 222)

The significance of *self-consciousness*, of people being aware of themselves, of people choosing how best to organize their lives so as to fulfil their potentials, was addressed in Chapter 14, and will not be repeated here. Animals that do not have the degree of social interdependence exhibited by humans, and have not genetically evolved the facility of self-consciousness, a mind, and the ability to communicate through symbolic representation, are at a disadvantage in passing on their learning to their offspring. Crucially, for humans, knowledge advances with experience, influencing choice and behaviour. Hence understanding activity must go deeper than reductionism, empirically challenging hypotheses that describe experience; or holism, testing the plausibility of a paradigm to theoretically explain the parameters of that experience.

Humans do more than test hypotheses against *what* they experience, or judge the relevance of a paradigm to explain *how* immediate experience is facilitated. We certainly live in communities shaped by cultural forces that influence our vision, the paradigm, the intellectual parameters of our experience. But people try, intuitively, to go beyond explaining the here and now, and assess *why* we behave as we do. People try to anticipate *what will come*. Effectively, humans second-guess the revolutionary science of the experts, the period of intellectual confusion when existing 'official' modes of thought are unable to account for experience. In periods of social change, when extant understandings of behaviour fail to anticipate people's activity, people do not run around like headless chickens waiting for experts to agree on a new model of reality. In such times people (and scientists) fall back on their intuitions, their beliefs, as to what is likely to/might happen: beliefs that evolve to understand experience, and which are an incipient theory of human nature.

The actual world of individual experience is a microcosm of the social real world, but not an exact replica. And the myriad of forces that impinge upon our experience contribute to our knowledge: the process of *praxis*. In this sense the future is radically unpredictable, and for us as individuals beyond our control, even though we will do our best to reduce uncertainty by trying to control the conditions of our existence by exerting power over other people. But this uncertainty allows us, socially, 'to construct *our own futures*, albeit in circumstances not of our own choosing' (Rose 1997: 307, emphasis added). The future is not determined by forces beyond our control; we do *not* have to be passive victims of genetic evolution or inexorable social forces. We can, potentially, purposefully intervene in social existence to guide the unfolding path of our experience.

But of course the first prerequisite of such a proactive response to existence is an awareness of the 'real' world beyond individual experience. As postmodernists argue, people *do* experience different, actual realities, but we also share a common existence: there *is* a real world beyond our own experience of that world.

> To put it formally, we live in a material world which is an ontological unity, but which we approach with epistemological diversity.
>
> (Rose 1997: 304)

The real world exists, but can *only* be perceived theoretically: perceptions that reflect real experience. This world cannot be observed, but is revealed through observation. To discover this world, different experiences have to be compared and contrasted, and the question asked: 'What must the world be like, for it to appear in so many different guises to so many different people?' In the case of the environment:

what continues to strike me is that the 'environmental issue' necessarily means such different things to different people . . .

(Harvey 1993: 1)

The comparison of distinct, empirically corroborated viewpoints, is essential in scientific enquiry. And when this real world is conceptually modelled, there is no vision of the real world untainted by methodological bias to which this perception can be compared. There is no reality check beyond individuals' differing experiences. And that reality is constantly changing, as people change their perceptions of that world in the light of their experience – the process of *praxis*. Reality is a process in constant change. It is complex, with

> many intertwining strands of cause and effect that combine, within some consistent world view, to constrain and control the unfolding of a particular selection of events . . . [Such a world view] sees not just one interpretation of reality, but many, yet it sees them as a seamless whole . . .
>
> (Stewart and Cohen 1997: 289)

Dialectics and understanding

So, what is this 'world view' that understands 'many interpretations of reality' as a 'seamless whole'?

Individual behaviour and social existence are understood to be radically indeterminate, and yet people can, potentially, construct their own futures in general, and in the context of this book, their economic, environmental and developmental futures in particular. People can choose and thereby make their own history. The indeterminacy of life does not imply ignorance; it is inherent in social existence. As a process, reality goes beyond the simple dichotomy of truth/falsity; it is a question of understanding individual experience in the context of changing social existence. Reality is a dialectic through which individuals can socially alter their own futures: an interpretation of experience in the light of individuals' social potentials. The process of social change is a combination of chance and necessity.

As individuals, people have to fulfil certain biological functions: eat, sleep, defecate, etc. As social animals people have to: reproduce, learn from each other, cooperate to produce, socially interact, etc. As a society people have to organize themselves: through institutions to live as social animals so as to collectively produce, in production processes that maintain the integrity of the natural environment to maintain human life, and through social change to be able to realize an improvement in livelihoods that better fulfil their potentials – and at the same time have sufficient social flexibility and personal confidence to make up their minds as to their preferred activity.

> The first premise of all human history is, of course, the existence of living human individuals. Thus the first fact to be established is the physical organization of these individuals and their consequent relation to the rest of nature . . . The writing of history must always set out from these natural bases and their modification in the course of history *through the action of men.*
>
> (Marx and Engels 1970: 42, emphasis added)

This is a problematic that implies a dialectical methodology. The properties of the parts – individuals – are not abstracted from their association in wholes – societies. Rather, individuals' characteristics are a consequence of their being members of societies, and societies reflect individuals' particular personalities, institutions and motivations. Neither is ontologically prior: variables cannot be catagorized as the independent subject (determinant, causal) or dependent object (determined, caused); both are the condition and the effect of the other. But dialectics implies *more than* interdependence. People creatively interact between themselves, and within social institutions, producing consequences that were often unintended by any of those involved, but of course in as far as people have power in society, they are more able, at least in the short term, to impose their choices and will on other people.

> For dialectics the universe is unitary but always in change; the phenomena we can see at any instant are parts of processes, processes with histories and futures whose paths are not uniquely determined by their constituent units. Wholes are composed of units whose properties may be described, but the interaction of these units in the construction of the whole generates complexities that result in products qualitatively different from the component parts . . . In a world in which such complex developmental interactions are always occurring, *history becomes of paramount importance . . . the past imposes contingencies on the present and the future.*
>
> (Rose *et al.* 1984: 11, emphasis added)

Formalizations of *the* dialectic impart a rigidity and dogmatism to the concept, which denies its fluidity in addressing historically specific contingencies. It is not a series of laws imposed by nature, analogous to the quantitative relations between phenomena in the natural sciences. Dialectical analyses do not predict anything, but as a methodology dialectics does highlight what is *possible*. Such analyses are not alternatives to orthodox scientific argument or logic, arguments and logics that are perfectly valid and coherent within their intellectual parameters, but which can only ask certain questions. 'The dialectic is not a calculator into which it is possible to push the problem and allow it to compute the solution' (Rees 1998: 271).

> Dialectics cannot be imposed on the facts; it has to be deduced from the facts, from their nature and development. Only painstaking work on

a vast amount of material enabled Marx to advance the dialectical system of economics to the conception of value as social labour.

(Trotsky 1973: 233)

No way of thinking can address all of the infinitely complex interactions between phenomena, and dialectics is similarly partial, addressing the potentials for change and a strategy for their possible realization. 'Potential' has no location; it is an emergent process with 'a shimmering haze of possibility' (Stewart and Cohen 1997: 303) in the material, 'real' world: a possibility contingent on an appropriate theorization of individual and social existence and on human motivation and social activity. The evolutionary process has to be reconstructed to identify the social construction of necessity, the parameters within which people are able to choose; and voluntaristic notions of chance, reflecting individuals' idiosyncratic preferred options. In what ways people are constrained to follow social priorities, and how these priorities might be adapted to meet the needs of creative individuals, is conceptualized and theorized.

> The first principle of a dialectical view, then, is that a whole is a relation of heterogeneous parts that have no prior independent existence *as parts*.
>
> (Levins and Lewontin 1985: 273, emphasis added)

The parts do not have intrinsic characteristics separable from the wholes of which they are a part.

> The second principle . . . is that, in general . . . the properties of the parts . . . are *acquired by being* parts of a particular whole.'
>
> (Levins and Lewontin 1985: 273, emphasis added)

The characteristics of the parts, and the wholes of which they are constituents, are not immutable. There is a constant interaction between the parts, and between the parts and the whole.

> A third dialectical principle, then, is that the interpenetration of parts and wholes is a consequence of the *interchangeability of subject and object*, cause and effect.
>
> (Levins and Lewontin 1985: 274, emphasis added)

No system is completely static, though in some aspects it may be in temporary equilibrium. Such stasis is not the natural order of things. In the dialectical vision of constant change, persistence has to be explained as a consequence of a particular balance of positive and negative feedbacks limiting the parameters for change. For instance, in Chapter 5, in the discussion on the tendency for the rate of profit to fall, it was noted that, for instance, if the rate of exploitation increases to offset the effect of investment in production which proportionately employs less labour than before, then, although there is a tendency for the rate of profit to fall, the *actual* rate of profit would not fall. And such a tendency explains the decisions that have to be made to maintain profitability, and resultant pressure to increase the rate of exploitation of labour, which will engender social conflict, possibly leading to social change. In this context, equilibrium is a balance of positive and negative feedbacks: it is not stasis.

Heterogeneous processes and forces interact to generate change. But in this sense heterogeny is more than diversity; processes and forces interact as opposites, and contradictions between them create change and a product qualitatively different from the interacting parts. The future cannot be predicted by extrapolating past trends. Complex processes – the ecology, economies, societies – change apparently spontaneously. They clearly respond to changes in other processes, but movement may not necessarily be the result of changes in external parameters. For instance, if economic competition functions properly it negates itself, producing monopoly, as the least efficient enterprises go out of business.

Material reality and individual choice

It is the indeterminacy of the interaction of co-determinant, creative wills, in a society in change that creates the confusion that has led postmodern theorists to give up trying to conceptualize the 'real' world.

> [I]t is impossible to simply stare at the world as it immediately presents itself to our eyes and hope to understand it. To make sense of the world we must bring to it a framework composed of elements of our past experience; what we have learned from others' experience, both in the present and in the past; and of our later reflections on and theories about this experience.
>
> (Rees 1998: 63)

Even the most abstract conceptualization and theorization of existence has its basis in *real* experience. These conceptualizations, human consciousness, are produced by creative humans trying to understand their existence so that they can purposefully choose how to better organize their affairs to fulfil their potentials. And the understanding of experience is mediated by beliefs, which rationalize and make sense of experience.

> Consciousness can never be anything else than conscious being, and the being of men is their conscious life process.
>
> (Marx and Engels 1970, quoted Rees 1998: 64)

Beliefs reflect individuals' experience, which is culturally filtered and interpreted to make individual actual experience commensurable to the wider, social, real world. Intellectuals call these beliefs 'assumptions', and it requires concentrated intellectual effort to identify the 'incompletely explicit assumptions, or more or less *unconscious mental habits* [beliefs] operating in the thought of an individual or a generation . . . [which define] the dominant *intellectual tendencies* of an age' (Lovejoy 1964: 7, quoted Harvey 1993, emphasis added).

'It is not the consciousness of men that determines their being, but on the contrary it is their social being that determines their consciousness' (Marx 1976b: 3). But of course individuals' minds affect their *own* social being, so consciousness, theory, becomes a co-determinant of experience and social being: within social parameters individuals *are* free to choose. Theory, then, becomes a material force in changing people's reality – but theory *alone* is insufficient: it is the theorization of real experience that is effective. Theory furnishes a conception of the real world, but this is a reality *always* in change and which is *never* fully appropriated by theory: 'man . . . possesses consciousness, but, even so, not inherent, not "pure" consciousness' (Marx and Engels 1970: 50). And if theory is to bring about intended change, purposeful activity has to be more than theoretically describing past events, or explaining extant parameters to individual behaviour.

> The philosophers have only interpreted the world in various ways; the point is to *change* it.
>
> (Marx, *Theses on Fuerbach*, 1975b: 423)

Change is motivated by people trying to fulfil their changing creative potentials. And theories are only one component of social change; material reality is the other. But if we can never *fully* theorize and know this material reality, what is our guide to this 'real world'?

In the last analysis, humans' ability to consciously change social activity towards a preferred way of life is based on changing ways of organizing human labour. People *create* their world through labour.

'. . .[T]he first premise of all human existence and, therefore, of all history . . . [is] that men must be in a position to live in order to be able to "make history"' (Marx and Engels 1970: 48). And what characterizes human existence is the production of the means of subsistence: tools, technology, ideas for the organization of production, etc. – production for an imagined contingency – a consequence of 'self-awareness'.

> A spider conducts operations that resemble those of a weaver, and a bee puts to shame many an architect in the construction of her cells. But what distinguishes the worst architect from the best of bees is this, that the architect raises the structure in the imagination of the labourer at its commencement. He not only effects a change of form but he also realizes a purpose of his own.
>
> (Marx 1974, quoted Ollman 1976: 111)

People self-consciously imagine what they will do. But, as we saw in Chapter 14, this mental capacity, the mind, evolved as humans, in the material world, produced their survival. 'Man has made his history . . . by creating through his labour an artificial environment by developing successively his technical aptitudes and by accumulating and transforming the products of his activity and this new environment' (Labriola 1904: 77, quoted Rees 1998: 264).

> The production of ideas . . . of consciousness, is . . . the material inter-
> course of men, the language of real life . . . the direct efflux of their
> material behaviour . . . Consciousness can never be anything else
> than conscious existence, and the existence of men in their *actual*
> life process.
>
> (Marx and Engels 1970: 47, emphasis added)

Consciousness is a consequence of people's material life, which in turn reflects their consciousness: the dialectic of social existence. But while conscious-ness reflects social existence, it is 'at first, of course, merely consciousness concerning the *immediate* sensuous environment and consciousness of the limited connection with other persons and things outside the individual' (Marx and Engels 1970: 51). People have to understand their *immediate* experience if they are to choose how to better organize their activity to fulfil their potentials.

Everybody has creative potentials, and a society that would facilitate the fulfilment of this creative energy would be one based on economic equality and political participation: *socialism*. Such a development priority was discussed in Chapter 13, and will not be repeated here. The present concern is to highlight the generation of knowledge towards such a social transformation.

Chapter 5 highlighted the analyses of economies as *processes*, where rela-tions of exchange between producers and consumers are relations between citizens. And citizens are fundamentally constrained in fulfilling their economic potentials, and therefore their income and their livelihoods, by the class nature of the capitalist mode of production. Capitalist class power is based on the control of the social means of production, implying contradictory class interests, a trend towards economic crisis, and social conflict. But people experience economic crisis and social conflict in different contexts: workers may become unemployed, or suffer low wages; students have insufficient grants to complete their studies; hospital waiting lists mean the sick go untreated, or treatment is withheld; teachers have to work with large class sizes, or there are insufficient numbers of schools and teachers; the poor suffer bad housing and poor environmental conditions, and in some coun-tries starve to death; etc. *All* of these constraints reflect *who* controls the social means of production. Because of people's experience, a '*general* struggle against capitalist forms of domination is always made up of *particular* strug-gles' (Harvey 1993: 44, emphasis in original).

The resolution of the 'particular' struggles of workers, students, patients, teachers, etc. will only be achieved if the underprivileged and disaffected in whatever walk of life combine in a general struggle to challenge the control of economic resources by a privileged minority. The problem is: how does this common class *interest* become a common *motive* behind individuals' behaviour and activity?

> [P]eople are not machines . . . there are subjective processes in the making of decisions . . . it is not classes but individuals who make choices . . . the consciousness of an individual is not determined by his or her class position . . .
>
> (Lewontin and Levins 1997b: 68)

Knowledge and change

Knowledge as *praxis* is a process by which people explain their own history and make it at the same time, linking chance and necessity, highlighting the social parameters of individual freedom. People become critically aware of their *own* experience: praxis 'draws out and elaborates that which people already "feel" but do not "know"' (Foracs 1988: 323). People develop their own perception of the world based upon their experience.

> We make and change the world only through the mind of man, through his will for work, his longing for happiness, in brief, through his psychological existence . . .
>
> (Gramsci, quoted Boggs 1976: 57)

The awareness of a shared class interest implies a conceptual understanding of modes of production, and in particular the *capitalist* mode of production: that the power of capital is enforced through commodity exchange. Through a social process of exploitation, surplus value becomes the unearned income of the ruling class, directing production towards the accumulation of capital rather than the satisfaction of human need. And the tendency towards economic crisis is inherent in the process of capitalist society, and if the interest of the proletariat is not defended the constraints on their social existence will only get worse. But rather than this theory merely being explicitly understood, it has also to become part of people's 'common sense'; part of their vision, their world view for understanding experience.

> . . . every philosophy has the tendency to become the common sense of a fairly limited environment (that of intellectuals). It is a matter of starting with a philosophy which . . . is connected to and implicit in practice and elaborating it so that it becomes a renewed common sense . . . through a critique of capitalist civilization . . . A critique implies . . . self-consciousness. To know oneself means . . . to be master of oneself, to distinguish

oneself, to free oneself from a state of chaos, to exist as an element of order – but of one's own order and one's own discipline in striving for an ideal.

(Gramsci 1916, in Foracs 1988: 332, 59)

People are thinking, creative, social beings who can readily participate in the process of cultural change, a process of changing common sense consequent on a changed consciousness, a different way of interpreting experience. And not only in terms of individuals' own experience: 'To make sense of the world, we must bring to it a framework composed of elements of past experience; what we have learned of others' experience, both in the present and in the past; and of our later reflections on and theories about this experience' (Rees 1998: 63). This process of change challenges the cultural hegemony of the ruling class, which disseminates ways of thinking through schools, universities, the mass media, etc. that are broadly supportive of the social status quo. The class struggle is also an ideological struggle.

> To the extent that ideologies are historically necessary they have a validity which is psychological, they 'organise' human masses, and create the terrain on which men move, acquire consciousness of their position, struggle, etc.
>
> (Hoare and Nowell Smith 1971: 367)

Ideas are not imposed mechanically from the outside by experts, but consciousness evolves as people become critically aware of their social experience. And here there is a role for intellectuals, people who have acquired the skills, have the time, and the requisite personal confidence to abstractly analyse social life. Typically intellectuals work in institutions that reinforce the cultural hegemony of the social status quo, and reproduce themselves through the educational system; criticism is allowed within more or less tight boundaries, and intellectuals are generally not encouraged (or allowed) to engage in *praxis*, putting critical ideas into practice.

> A human mass does not 'distinguish' itself, does not become independent in its own right without, in the widest sense, organizing itself; and there is no organization without intellectuals, that is without organizers and leaders, in other words, without the theoretical aspect of the theory–practice nexus being distinguished concretely by the existence of a groups of people 'specialized' in conceptual and philosophical elaboration of ideas ... The process of development is tied to a dialectic between the intellectuals and the masses.
>
> (Gramsci, in Foracs 1988: 334–5)

The intellectual is not now the expert, but an activist in the process of social change. And knowledge reflects experience. Development and progress is 'all about liberating people from all that holds them back from a full human life' (Hope and Trimel 1995: 9).

Seeing ourselves through ourselves

Because humans are self-conscious they can know themselves. Because people can know themselves they can be aware of their social potentials. Because humans can be aware of their social potentials they can choose how to organize their lives to better fulfil their potentials: *development*.

For humans to know themselves the generation of knowledge must reflect their experience. And only people themselves can recount their social activity, reflecting their intuitions and what their potentials might be; their actual experience. Individuals' actual experience, to be understood, has to be situated and conceived of as a facet of much broader, complex, and inclusive social relationships. Such a contextual appreciation can only be a consequence of being empowered to share in the control of social existence: *participation*.

Relevant knowledge is about people better understanding the social significance of their individual experience. Such knowledge allows individuals to adapt their behaviour to realize ever more of their changing, social potentials: *praxis*.

Three conceptions – development, participation, praxis – one meaning: the fulfilment of the creative potentials of human beings.

> I must, being within myself, see myself through myself.
> (José Martí, the Cuban nineteenth century revolutionary, quoted Cardoso 1997: 170)

The politics of knowledge

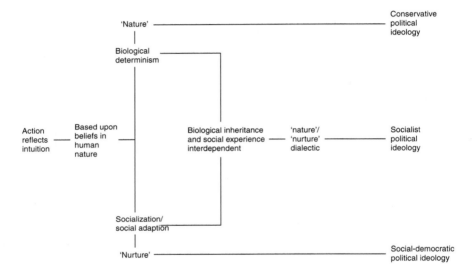

18 Intellectual panorama, ideological vision, and political view

Conceptualizing reality

> There is an infinity of facts that we could have but only a small selection available. The selection exists because we have particular assumptions about the world, theories . . . It is part of the conventional wisdom that you should believe the evidence of your eyes . . . but in our world you cannot see the most important things, they are abstractions, concepts.
>
> (Harris 1983: 7, 10)

To be able to change the world we have to understand it. Understanding requires that we conceptualize experience, both our own, and others' vicariously through reading. But as we have seen in Parts I-IV, key aspects of our lives – economic relations, the way we interact with the environment, processes of change and development – can all be coherently conceptualized very differently. The real world is conceived in distinct ways.

Theories that purport to understand human behaviour can furnish no adequate answers unless, at least implicitly, they are based on a conception of what it is to be human. And to try to contrast and compare, and to learn from the insights of the plethora of schools of thought on social activity, I have found it necessary to categorize theoretical analyses by reference to the unexplained and unexamined assumptions about human nature which define the theoretical world view – the 'set of preconceptions that provide the framework for . . . [the] analysis of the world' (Levins and Lewontin 1985: 267). Strictly speaking, these are not the assumptions of a theory, but the reason *why* particular assumptions are chosen. A world view sets the beliefs that define the 'good society', specify the meaning of 'justice', distinguish between right and wrong, and set the standards for what men and women 'should strive to be'.

In this context, people as social beings can be conceived of as: instinctively acting according to their biological make-up – *nature*; or responding to cooperative social pressures to survive – *nurture*; or discovering their creative potentials through social experience – *nature–nurture dialectic*. Such beliefs about human nature are not amenable to empirical verification, since such a priori beliefs define the evidence by which the relevance of the theory would be demonstrated. The argument is circular: it *is* a question of beliefs.

Cole's Law

Although this book has been concerned to explain the significance of alternative perspectives in social analysis in general, and economic, environmental theory and development studies in particular, such a tripartite approach characterizing theoretical perspectives is very common (see Table 18.1).

Ever since my days as an undergraduate, when I first became aware of different theoretical approaches to defining reality, I have been trying to reconcile distinct accounts of essentially the *same* experience. As a student I worked hard, as at that time the alternative for me was working in a car factory in Birmingham. I felt relieved and satisfied after my last finals exam, and that evening sat down to relax and unwind watching television. And I was particularly interested to see a programme featuring different economists discussing the long-term economic decline of Britain. I thought that by now I had mastered the principles of economics and the fundamentals of development studies, and would be able to understand the theoretical nuances of explanations of economic decline. The programme featured three economists offering explanations of the lacklustre performance of the British economy, and perhaps arrogantly I expected after my years of study, reading for a degree in development studies (specializing in economics), it would be almost like a 'conversation between equals'. However, of the three experts I could only really understand the analysis of one – which I now realize was because he understood the issue in terms of the implicit beliefs that underlay my university course in development economics. I was tempted to dismiss the other two as incompetent, but they were senior academics and government advisors: authorities on economic decline. But at least I had the consolation that it was evident that the three presenters were similarly at a loss to understand each other!

An awareness of theoretical confusion has remained at the back of my mind ever since, and over the past twenty years or so, as I have been preparing teaching materials, conference papers, manuscripts for articles and books, reports, etc. I have kept a note of examples of alternative perceptions of the same experience: different conceptions of 'reality'. Table 18.1 cites over 170 examples, drawn from writings on economics, ecology, politics, biology, sociology, philosophy, geography, and development studies, on such issues as trade, economic integration, ideology, social change, science, evolution, disability, human nature, gender, social welfare, debt, employment and industrialization, with case studies from South Africa, Cuba, Namibia, the Caribbean, Great Britain, Africa, United State of America, Europe and South America. And I have grouped these categorizations into three columns, reflecting the (usually) implicit beliefs about human nature that rationalize the various analyses.

Table 18.1 Tripartite characterizations of social analyses

	I	II	III
Adams (1995)	individualist	hierarchists	egalitarians
Addis et al. (1992)	libertarianism	pluralism	Marxism
Alford and Friedland (1985)	managerial	pluralist	class
Allen and Thomas (1992)	neo-liberalism	interventionism/populism	structuralism
Allison (1971)	rational actor	organizational process	governmental politics
Appelbaum (1976)	equilibrium	evolutionary	conflict
Arnstein (1969)	non-participation	tokenism	degrees of power
Atkinson (1989)	supply side/free market	Keynesian/interventionism	Marxist
Atkinson (1991)	intellectual crisis	environmental crisis	social crisis
Axline (1979)	laissez-faire	dirigiste	dependency-reducing
Backhaw (1992)	subject	object	subject/object
Barbier (1989)	descriptive	functional	evolutionary
Bardham (1989)	neoclassical	structuralist-institutionalist	Marxist
Barrios Villegas (1987)	parastatal NGOs	professional NGOs	politically progressive NGOs
Barry-Jones (1983)	liberal	national political economy	Marxist
Beardesley (1981)	liberal	pluralist	Marxist
Benton (1977)	positivism	humanism	historical materialism
Benton (1993)	biocentrism	ecocentrism	anthropocentrism
Berry (1983)	conservative	liberal	Marxist
Bhaskar (1979)	Weber	Durkheim	Marx
Biot et al. (1991)	neo-liberal	classic	populist
Blaikie (1995)	neo-liberal	paternalist	populist
Blowers (1984)	neo-elitism	pluralism	Marxist structuralism
Bottomore (1985)	Weber	Schumpeter	Marx
Boulding (1989)	economic power	threat power	integrative power
Brett (1985)	orthodox	reformist	Marxist
Brown (1974)	classical	Keynesian	Marxist
Cameron (1984)	reactionary	reformist	innovatory
Chenery (1975)	neoclassical	structuralist	neo-Marxist

Table 18.1 Continued

	I	II	III
Coates and Hillard (1986)	right	centre	Marxist
Colclough (1982)	neoclassical	structuralist	neo-Marxist
Cole (1983)	individualist	technocratic	collectivist
Cole (1987)	subjective preference theory	cost-of-production theory	abstract labour theory
Cole (1994)	independent individuals	dependent individuals	interdependent individuals
Cole (1995)	consumer	producer	citizen
Cole et al. (1991)	subjective preference theory	cost-of-production theory	abstract labour theory
Corbridge (1986)	'new' neoclassical	Keynesian/interventionist	radical/world systems
Cox et al. (1985)	elite	pluralist	Marxist
Cox (1981)	neo-realism	hegemonic stability	hegemonic change
Cox (1993)	liberal democratic	producer self-management	popular participation
Crook et al. (1992)	Weberian	Durkheimian	Marxian
David (1986)	orthodox paradigm	mid-range perspectives	radical heterodox paradigm
Desai (1974)	neoclassical value theory	classical value theory	Marxian value theory
de Santamaria (1992)	subjective	objective	collective participation
Dryzek and Lester (1989)	conservative environmentalism	environmentalist centre	radical environmentalists
Duncan (1983)	conservative	liberal	radical
Elshtain (1985)	Hobbesian feminism	cultural feminism	feminism as citizenship
Espring-Anderson (1990)	conservative	liberal	Marxist
Evans (1989)	neoclassical	neo-Ricardian	Marxian
Fine (1990)	free market	technical needs of capital	class struggle
Frank (1996)	neoclassical (right)	Keynesian (centre)	Marxist (left)
Friedmann (1987)	conservative	reformist	radical
Fyson (1984)	alternative	liberal	radical
Gill and Law (1989)	liberalism	liberal	Marxism
Gilpin (1975)	liberalism	realist-mercantilist	Marxism
Glass and Johnson (1989)	falsificationism	mercantilism	instrumentalism
Goldstein (1988)	conservative	inductivism	revolutionary
Goldstone (1986)	modernization and revolution	structural approach to revolution	classic approach to revolution

Table 18.1 Continued

	I	II	III
	positivism	realism	structuralism
Keat and Urry (1982)			
Kindleberger and Herrick (1977)	neoclassical	structural	radical
Kirkpatrick (1987)	neoclassical	structuralist	radical
Kitromilides (1985)	technocratic	pluralist	Marxist
Kristol (1981)	neo-Austrian school	post-Keynesian	radical-humanistic economics
Kuttner (1985)	neoclassicism	post-Keynesian	neo-Marxism
Lal (1976)	neoclassicism	structuralist	dominance/dependency
Lappe and Schurman (1989)	biological determinist perspective	descriptive social perspective	power structure perspective
Lawson (1992)	induction	deduction	realist
Layder (1994)	individual and society	agency-structure	macro–micro
Lee and Newby (1987)	community as location	community as system	community as relationship
le Roux (1990)	free market	social democratic	socialist
LeShan and Margeneau (1982)	one-dimensional reality	two-dimensional reality	three-dimensional reality
Levins and Lewontin (1985)	conservative	radical	revolutionary
Levins and Lewontin (1994)	reductionism	holism	dialectics
Lisk (1977)	growth oriented	employment oriented	poverty oriented
Little and Smith (1991)	politics of power and security	politics of interdependence	politics of dominance
Lukes (1977)	elitist	pluralist	Marxism
Lutz (1992)	mainstream	humanistic	Marxist
Mackintosh (1992)	policy as private interest	policy as prescription	policy as process
Magnum *et al.* (1987)	neoclassicism	institutionalism	radicalism
McGrath (1990)	liberal orthodoxy	political reformist	Marxist revisionist
McKinley and Little (1986)	liberal	realist	socialist
Mannathoko (1992)	liberal	radical	socialist
Marshall (1987)	temporary recession	structural shock	crisis of capitalism
Mellos (1988)	neo-Malthusianism	eco-development	radical ecology
Merchant (1992)	self: egocentric	cosmos: ecocentric	society: homocentric
Miles (1995)	nature/sex	culture/gender	nature–culture/historical category
Mohun (1989)	market optimists	market pessimists	Marxists

Table 18.1 Continued

	I	II	III
Morrison (1995)	Weber	Durkheim	Marx
Mubende (1986)	modernization theory	underdevelopment theory	social formation theory
Oliver (1996)	personal tragedy	sociopolitical	political economy
Pateman (1970)	psuedo-participation	partial participation	full participation
Pearce (1989)	knowledge based	procedure based	needs based
Pearce and Turner (1990)	bio-determinism	cultural determinism	organism and its environment interact
Peet and Watts (1996b)	market	state	civilization
Pepper (1993)	independent development	dependent development	articulated development
Pepper (1996)	elitism	pluralism	Marxism
Petras (1978)	liberal	structural	radical
Pilgrim and Rogers (1994)	the user as patient	the user as consumer	the user as survivor
Pourgerami (1990)	humanist	institutionalist	dependence
Preston (1982)	positivist	radical	Marxist
Reason and Rowan (1981)	naive inquiry	old paradigm research	new paradigm research
Redclift (1987)	Weber/neoclassical economics	Durkheim/deep ecological positions	Marx/Marxist theory
Reynolds (1971)	capitalist	less developed	socialist
Ritzer (1975)	social definition	social facts	social behaviour
Robinson and Williamson (1983)	neoclassical	post-Keynesian	Marxist
Romano and Leiman (1970)	conservative	liberal	radical
Rose (1973)	reductionist	holistic	interactive
Rowthorn (1974)	neoclassical	neo-Ricardian	Marxist
Russett and Starr (1996)	realists	traditionalists	radicals
Saul (1994)	liberal democracy	authoritarianism	democratic empowerment
Seidman and Anang (1992)	mainstream	basic needs/structuralist	transforming institutionalist
Seitz (1988)	orthodox	growth of equity	radical
Seitz (1995)	market approach	state approach	civil society approach
Sender and Smith (1988)	free market	imperialist restrictions	class power
Shaw and Gouldery (1982)	conservative	reformist	superstructural
Smith (1983)	conservative	liberal	Marxist

Table 18.1 Continued

	I	II	III
Smith and Toye (1979)	evolutionist	technocratic	Marxist
So (1990)	modernization	dependency	world system
Srinivasan (1989)	neoclassical	structuralist	Marxian
Staniland (1985)	behavioural	interactive	deterministic
Stankiewicz (1992)	conservatism	liberalism	socialism
Stewart and Streeten (1976)	price mechanisms	technologists	radicals
Stiglitz (1988)	neoclassical/rational peasant model	information-theoretical approach	institutional-historic approach
Stiles (1990)	functional	neo-functionalist	political dominance
Stinchcombe (1978)	narrative sequences	causal structures	epochal interpretations
Stone (1986)	authoritarianism	democratic pluralist	populist-statist
Strange (1988)	national manufacturing base	transnational military-industrial complex	international hegemony
Streeten et al. (1981)	conservative basic needs	reformist basic needs	revolutionary basic needs
Toye (1986)	neoclassical	classical political	political economy
Turner (1993)	absolute scarcity	relative scarcity	social limits
Turner (1995)	neoclassical	ecological	humanistic
UNIN (1986)	incremental	structural	revolutionary
Vaillancourt (1992)	conservationist	environmentalist	political-ecologist
Varoufakis and Young (1990)	Austrian	neoclassical	Marxian
Wager (1991)	techno-liberals	counter-culturalists	radicals
Wallerstein (1980)	conservative	liberal	Marxist
Walters and Black (1992)	liberal	mercantilist	Marxist
Ward (1979)	conservative	liberal	radical
Weiss (1983)	neoclassical	structuralists	radical
Wheelock (1992)	orthodox economics	total economic activity	Marxist economics
Williams (1994)	liberalism	realism/mercantilism	Marxist
Wilson (1992)	commutative justice	productive justice	distributive justice
Woodhouse (1992)	neo-liberal	institutional	populist
Yaxley (1996)	libertarian mechanism	egalitarian system	social process
Zohar and Marshall (1994)	society of individuals	society is disciplined army	society is free-form dance company

Column I refers to analyses that understand behaviour to reflect humans' innate biological instincts and potentials – *nature*. Because people are conceived of as 'genetically determined', human nature is essentially *pre-social*, and individuals' potentials are *independent* of society, society being merely the sum of the individuals of which it is composed.

The analyses listed under column II see social existence as reflecting the need for people to cooperate. Fundamentally, human potentials reflect individuals' abilities to work together essentially with regard to a technical division of labour. People specialize in social and economic activity. And people are socialized by society to fulfil roles defined by society – *nurture*. Individuals are no longer independent *of* society, but dependent *on* society: society is more than the sum of the individuals of which it is composed.

In column III the analyses appear to presume that although people possess innate attributes, these potentials can *only* be realized through society. People as creative beings discover their potentials through social experience; people *change* as a consequence of social life: changes that will crucially reflect the opportunities available in society to develop their potentials – a *nature–nurture dialectic* of justice. People are not independent of, nor dependent on, but interdependent *within* society. Society is not only more than the sum of individuals; it also has a structure of inequality that limits the opportunities of the powerless.

All human existence is social existence, and it can never be obvious what is the effect of biology and society on individual behaviour. One or another of these conceptions of individuals' motivation, beliefs in human nature, underlies *all* analyses of human behaviour.

Hence Cole's Law:

> *In the understanding of human activity there are always coherent, alternative explanations, which fundamentally reflect different beliefs in human nature.*

Ideology and intellectual panoramas

Different intellectual disciplines and schools of thought within the same world view or perspective share intellectual parameters that define the real world. For instance, 'sociobiological analysis in the hands of E.O. Wilson and others employs identical mathematical models to those used by a particular school of monetarist economics based in Chicago' (Rose 1997: 53).

In Table 18.1, different authors addressing distinct issues, even though they are working within common perspectives (which are categorized in the same column), describe the different schools of thought using different terms. Theoretical perspectives, based upon a world view of human motivation (human nature), when applied to the analysis of actual experience, give rise to a number of schools of thought, which are essentially variations on a theme. For example, a *nature*-based approach to human nature is reflected in the subjective preference theory of value perspective in economics, which when applied to the analysis of inflation gives rise to a school of thought called *monetarism*. This school of thought shares a theoretical affinity with structural adjustment theory and new institutional economics, schools of thought that are derivatives of the modernization perspective, a consequence of the nature world view's being applied to development studies. Referring to new institutional economics and modernization theory, Geoffrey Hodgson writes that there is a 'consistency between socio-economic and biotic levels of analysis . . . [establishing] an important link between the social-economic and the natural world' (Hodgson 1998: 189). And that link is *ideological*.

The belief that underlies the *nature* world view as the foundation to social analyses is that behaviour is essentially a reflection of individuals' genetic inheritance. People are independent of society, and the emphasis in social policy is on individual freedom, so that individuals can fulfil their biological potentials. Other than in terms of equality of opportunity, issues of social inequality are outside the intellectual parameters of this perspective, and such questions cannot even be asked. The preferred regime to regulate social life is a policy of individual freedom and free exchange aimed at moving towards perfect competition, backed up by a strong state to guarantee these free-doms through the legal system with such statutes as the law of contract (see Chapter 11).

In contradistinction, a *nurture* world view defines theoretical perspectives that emphasize the need for cooperation rather than competition between individuals. As an ideology, technical expertise is justified, and since social life is a consequence of cooperation between people, extremes of wealth and poverty are considered to be dysfunctional, creating social strife and disorder rather than being incentives to individuals to maximize their potentials (see Chapter 12).

Where human nature is considered to be the result of a *dialectic between nature and nurture*, between biological endowment and social experience, then individuals are conceived of as creative beings, who evolve with the opportunities afforded by social experience to expand their potentials. And socially, everybody can best fulfil their potentials in as far as everyone else is similarly able to make the most of their innate attributes through social activity. It is not that societies have to be managed by experts to fulfil poten-tials, but that people have to participate in the control of society if they are to create and take advantage of the social opportunities most appropriate to their genetic potential. But social participation is a consequence of the development of class-consciousness: people becoming aware of their inter-dependence through class struggle (see Chapter 13).

That the same experience can be coherently analysed by different schools of thought, reflecting the emphases of distinct perspectives, expressing the moral emphases of different world views, is not a justification for theoretical relativism. The world view that underlies theoretical perspectives and schools of thought defines the moral and ideological parameters of the analysis of human experience: parameters that reflect different conceptions of 'knowledge' (as outlined in Part IV). Knowledge *itself* has a moral dimension. And the moral codes adopted by different people will reflect how they understand and make sense of their social experience. While theories may be *intellectually* coherent, for the individual they may not be *morally* coherent. People have to be aware of their own interests (and hence of their morality), based on a rationalization of their experience, and adopt a theoretical understanding of that experience appropriate to their moral sensibilities.

People all the time intuitively make up their minds as to where they stand on the issues in social life, but such choices only contribute to the emergence of class-consciousness, and purposeful participation in social organization, in as far as people are aware of the social implications of their individual behaviour. People's ability to understand *their* experience and needs will reflect the extent to which they are aware of and can rationalize *other* people's experience and perceptions of reality. Remember, the *real world* is a theoretical construction, which cannot be observed but is derived through observation: different people's perceptions of reality are compared and contrasted to determine what this shared world must be like for so many different perceptions to be generated.

In this sense the nature–nurture dialectic approach to understanding is inclusive of the two perspectives.

> Part of the work of the dialectician is ... to translate and transform other bodies of knowledge accumulated by different structures of enquiry and to show how such transformation and translations are revealing of new and often interesting insights ... dialectical ... materialist theory ... deals with totalities, particularities, motion, and fixity in a certain way, [and] holds out the prospect of embracing many other forms of theorizing within its frame ... There is a deep ontological principle involved here, for dialecticians ... elements, things, structures, and systems do not exist outside of and prior to the processes, flows, and relations that create, sustain, or undermine them.
>
> (Harvey 1996: 7, 9, 49)

The intellectual and society

Where the intellectual concern is merely to establish the sequence of events empirically, and theories are deemed 'scientific' as far as future events are successfully foreseen – *positivism* – then the intellectual is essentially an empiricist. Scientific work is geared to testing theories as empirically accurate representations of reality, and establishing statistical relationships between phenomena as the basis for future predication. Only empirical reality can be addressed, and only 'what?' questions can be asked (see Chapter 15). By not considering the abstract world of relationships, and not being able to ask 'how?' or 'why?' questions, essentially the status quo is accepted.

The approach is politically *conservative*.

When the scientific endeavour is defined as explaining the systematic relations between phenomena, certainly theories have to account for empirical reality, but science is more than this. Phenomena are not related through statistical relationships but through causative mechanisms, which are not in themselves empirically observable. Causative processes cannot be seen; causation is established by interpreting the *effects* of systemic relationships. Paradigms are devised to explain 'how?' empirical reality came to pass (see Chapter 16). The scientific status of a paradigm is now a reflection not of successful prediction, but of plausible argument. And when a paradigm fails to explain systems and cannot be relied on to solve problems then a period of revolutionary science redefines 'reality' until some predictability is restored to the understanding of experience: an experience that is shared, and managed in the general interest.

The approach is politically *social democratic*.

Praxis theorists not only attempt to ascertain empirical reality – 'what' happens – and to conceptualize systemic relations between phenomena – asking 'how?' questions – but try to understand, given all the possible interactions that *could* happen, 'why' that particular sequence of events was manifest, and what are the potentials for the future. Future potentials will differ between people, reflecting their consciousness of what is possible; which in turn will reflect the extent to which people are aware of shared interests and are able to organize themselves to prosecute such potentials. For praxis theorists such an awareness will understand the extent to which people's potentials are constrained by privilege consequent upon class power (see Chapter 17).

The approach is politically *socialist*.

Hence, is the intellectual a conservative empiricist, a social democratic expert, or a socialist activist?

The politics of knowledge

To what extent are intellectuals defined by their approach?

Often intellectuals treat different intellectual approaches are just so many techniques, to be kept in the academic toolbox to be used as appropriate to discover *the* truth. For instance, some may admit to being empiricists but claim to have socialist pretentions. And similarly some may see themselves as technocratic experts, but see some form of cooperative socialist future as desirable if not inevitable. The argument of this book has been that the achievement of social ends is contingent upon modes of thought as much as upon objective activity and subjective intent.

Logically, *if* human nature is believed to be independent of society and individuals necessarily act to further their own interests (however defined), *then* human potentials are best fulfilled in a competitive environment. And inequality is a consequence of individuals' 'natural' inadequacies. Such a legitimization of privilege has historically justified conservative political regimes and representative democracy.

But *if* individuals depend on society, and have to adapt to the needs of society, *then* potentials can best be realized through the expert management of social existence in the general interest. A meritocratic social order is legitimized, organized according to social democratic principles of fairness, within *pluralist* democratic institutions.

However, *if* humans are conceived of as a product of the dialectic between biological nature and social experience, *then* potentials are best achieved where people empower themselves to share in the control of society, and people are interdependent. A socialist organization of society is justified, organized according to the class interest of the powerless, through a process of *participative* democracy.

A socialist political agenda crucially reflects knowledge oriented to people's participation – *praxis*. People understanding themselves is a prerequisite for social participation – for which an empiricist, positivist approach to knowledge is inadequate, as is the emphasis on expertise and paradigms.

There is nothing wrong with being a positivist, a paradigm theorist, or attempting to engage in praxis. But there are logical implications to scientific method and the definition of knowledge, which limit the questions that *can* be asked: constraints that cannot be removed by 'wanting' to find different answers.

Positivist empiricism or paradigmatic systems approaches to scientific inquiry cannot develop the understanding of 'process' that is fundamental to the class-consciousness that underlies effective participation in the control of society.

Epilogue
What *do* we know about the world?

The process of reality and progress

The argument in all books is constrained by the preconceptions of the author. And this book is no different. The Prologue was an attempt to inform the reader of, as far as I am aware, the unwritten agenda, the vision, that this book expounds in discussing analyses and understandings of the economy, the environment, development and knowledge: an agenda that highlights the impossibility of considering human experience without acknowledging that the 'real world' exists beyond the individual. And this world can only be appropriated theoretically if analysts take cognizance that *all* conceptions of human existence are more or less biased by the experiences that inform understandings of that world.

Reality as a *process* becomes progressively inclusive as more and more individuals' experience is rationalized and articulated through the understanding of reality: an objective that is exacerbated as the creative essence of people's psyche evolves with social experience, and individuals realize new potentials, and thereby modify and transform their perceptions of experience, and hence of the 'real world'.

In this intellectual endeavour progress implies that intellectuals in general – and we are *all* intellectuals in one way or another – and professional intellectuals in particular – academics – increasingly address the multiple realities of social experience in their attempts to understand and theorize human experience. But such a cultural enlightenment appears ever less possible. As the world slides into economic recession, social conflict and moral confusion in the closing years of the second millennium, bigotry, self-interest, xenophobia, and intellectual myopia rather than enlightenment are becoming the staple intellectual diet. Paradoxically, the confusion and uncertainty amongst professional intellectuals that has spawned postmodernist approaches to knowledge has accentuated such reactionary trends in the ivory towers of universities. Robert Kuttner, reviewing trends in economic thought, comments:

> . . . there are two quite opposite trends in economics. One is centrifugal: there is a flowering of epistemological doubt . . . At the same time, there is a strong centripetal impulse. Economics *as practised* is unchanged, and

the resistance to real diversity within faculty ranks and classroom curricula is fiercer than ever. The debates are for the college of cardinals, not for the parish flock.

(Kuttner 1985: 83, emphasis in original)

In my days as a student, and in the early years of my work as an academic, I laboured under the illusion that the *raison d'être* of a university education was to conceptualize experience ever more accurately, adapting and improving the evolving theories of existence to encompass the nature of that changing experience more accurately. But it increasingly became apparent that academics in their pursuit of career advantage and kudos amongst their peers were averse to possibly controversial original thought and theoretical innovation. There has to be a return on their investment in intellectual capital. The mental habits and theoretical parameters that moulded their minds ten, twenty or thirty years ago are repeated in analyses of contemporary experience without taking into account the change in people consequent upon that experience. And the process of people's becoming more self-conscious with human development means that academic thinking and instruction becomes ever more anachronistic as experience is rationalized by the cultural priorities of yesterday.

Rather than seeking to expand knowledge by addressing an ever wider spectrum of changing human experience, many thinkers only 'know' what they were taught, with no expectation or motivation to extend and improve their understanding. Typically, educational institutions do not acknowledge or respect students' creative potentials to think for themselves. The old dogmas are presented as 'science' year after year, students being the 'disciples' expected to learn at the feet of the 'master'. Education is not geared to expanding intellects for people to become more aware of themselves, to improve the understanding of *their* experience to facilitate choice in how they will approach the future, through appreciating others' lives better. Creativity is not high on the intellectual agenda. Time-honoured truths are to be learnt and repeated, and if such conceptions fail to account for experience, then it is people's perception of that experience that is at fault, not the theory or theorist that is wanting.

I hope that, in some small way, this book will help to place creativity and self-expression somewhat higher on the cultural and educational agenda. In the ongoing process by which people, to effect progressive change, have constantly to free their minds from the constraints of past certainties that threaten an inevitable future, to embrace a changing present and the realization of new potentials yet to come, the theorization of the real world must become ever more inclusive of the experience of the disadvantaged, whose creativity and humanity are denied in order to preserve an iniquitous status quo.

The deepening contradictions of the world capitalist mode of production render privilege ever more tenuous, and the defence of class prerogatives becomes ever more ruthless, as competitive advantage sacrifices the potentials of the majority to the advantage of the few.

References

Abendroth, W. (1972) *A Short History of the European Working Class*, London: New Left Books.

Adams, J. (1995) *Risk*, London: UCL Press.

Addis, A., Selassie, B. and Seidman, R. (1992) 'On research on the state, law and legal processes of development, in A. Seidman and F. Anang (eds) *Towards a New Vision of Self-Sustainable Development*, Trenton, NJ: Africa World Press.

Alford, R. and Friedland (1985) *Powers of Theory*, Cambridge: Cambridge University Press.

Alfred, T. and Hennelly, S.J. (1992) *Liberation Theology: A Documentary History*, New York: Orbis.

Ali, T. and Watkins, S. (1998) *1968: Marching in the Streets*, London: Bloomsbury.

Allison, G. (1971) *Essence and Decision: Explaining the Cuban Missile Crisis*, Boston: Little Brown.

Amin, S. (1976) *Unequal Exchange*, New York: Monthly Review Press.

Anderson, T. and Leal, D. (1991) *Free Market Environmentalism*, San Francisco: Pacific Research Institute for Public Policy.

Andor, L. and Summers, M. (1998) *Market Failure: Eastern Europe's 'Economic Miracle'*, London: Pluto Press.

Appelbaum, R. (1976) *Theories of Social Change*, Chicago: Rand McNally.

Arestis, P. (ed.) (1988) *Post Keynesian Monetary Theory: New Approaches to Financial Modeling*, Aldershot: Edward Elgar.

Arnstein, S. (1969) 'A ladder of citizen participation', *Journal of American Institute of Planners*, July.

Atkinson, A. (1991) *Principles of Political Ecology*, London: Belhaven Press.

Atkinson, G. (1989) *Economics: Themes and Perspectives*, Ormskirk: Causeway Press.

Axline W. (1979) *Caribbean Integration*, London: Pinter.

Backhaw, H. (1992) 'Between philosophy and science: Marxian social economy as critical theory', in W. Bonefeld, R. Gunn and K. Psychopedis (eds) *Open Marxism: Volume 1 – Politics and History*, London: Pluto Press.

Backhouse, R.E. (1998) *Explorations in Economic Methodology: From Lakatos to Empirical Philosophy of Science*, London: Routledge.

Baker, T. (ed.) (1975) *The Long March of Everyman*, London: Deutsch.

Baldwin, R. (1995) 'Does sustainabilty require growth', in I. Goldin and L. Winters (eds) (1995) *The Economics of Sustainable Development*, Cambridge: Cambridge University Press.

Baran, P. (1967) *The Political Economy of Growth*, New York: Monthly Review Press.

Barbier, E. (1989) *Economics, Natural Resource Scarcity and Development*, London: Earthscan.

Barde, J.-P. and Pearce, D.W. (1991) *Valuing the Environment*, London: Earthscan.

Bardham, P. (1989) 'Alternative approaches to development economics', in H. Chenery and T. Srinivasan (eds) *Handbook of Development Economics*, Amsterdam: North Holland.

Barker, C. (ed.) (1987) *Revolutionary Rehearsals*, London: Bookmarks.

Barrios Villegas, F. (1987) 'ONGs y realidad nacional', in UNITAS, *El Rol de los ONGs en Bolivia*, La Paz: UNITAS.

Barry-Jones, R. (1983) *Perspectives on Political Economy*, London: Pinter.

Bartlett, R.V. (1989) 'Impact assessment as a policy strategy', in R.V. Bartlett (ed.) *Policy Through Impact Assessment*, New York: Greenwood Press.

Bauer, P.T. (1976) *Dissent on Development*, London: Weidenfeld & Nicolson.

Bauer, P.T. (1991) *The Development Frontier*, London: Harvester-Wheatsheaf.

Baumol, W.J. and Blinder, A.S. (1991) *Economics: Principles and Policy*, London: Harcourt Brace Jovanovich.

Beardesley, P. (1981) *Conflicting Ideologies in Political Economy*, London: Sage.

Beckerman, W. (1974) *In Defence of Economic Growth*, London: Jonathan Cape.

Beckles, H. and Shepherd, V. (1996) *Caribbean Freedom*, Princeton: Weiner.

Begg, D., Fischer, S. and Dornbusch, R. (1991) *Economics*, London: McGraw-Hill.

Beinart, W. and Coates, P. (1995) *Environmentalism and History*, London: Routledge.

Bell, D. (1981) 'Models and reality in economic discourse' in D. Bell and I. Kristol (eds) *The Crisis in Economic Theory*, New York: Basic Books.

Bell, D. and Kristol, I. (eds) (1981) *The Crisis in Economic Theory*, New York: Basic Books.

Bell, P. and Cleaver, H. (1982) 'Marx's crisis theory as a theory of class struggle', *Research in Political Economy* Vol. 5.

Benjamin, M., Collins, J. and Scott, M. (1994) *No Free Lunch: Revolution in Cuba Today*, New York: Grove Press.

Benton, T. (1977) *Philosophical Foundations of the Three Sociologies*, London: Routledge.

Benton, T. (1989) 'Marxism and natural limits: an ecological critique and reconstruction', *New Left Review* No. 178.

Benton, T. (1993) *Natural Relations*, London: Verso.

Bernal, J.D. (1969) *Science in History: Volume 4 – The Social Sciences – Conclusion*, Harmondsworth: Penguin.

Berry, C. (1983) 'Conservatism and human nature', in I. Forbes and S. Smith (eds) *Politics and Human Nature*, London: Frances Pinter.

Bettancourt, R. (1991) 'The new institutional economics and the study of the Cuban economy', in ASCE, *Cuba in Transition*, papers and proceedings of the first annual meeting of the Association for the Study of the Cuban Economy, Florida International University, Miami.

Bhaskar, R. (1979) *The Possibility of Naturalism*, Brighton: Harvester.

Biot, Y., Blaikie, P., Jackson, C. and Palmer-Jones, R. (1991) *Rethinking Research on Land Degradation in Developing Countries*, Washington: World Bank.

Birchall, J. (1987) '1968: all power to the imagination', in C. Barker (ed.) *Revolutionary Rehearsals*, London: Bookmarks.

Binns, P. and González, M. (1980) 'Cuba, Castro, socialism', *International Socialism* No. 28.

Blackburn, R. (1963) 'Prologue to the Cuban Revolution', *New Left Review* No. 21.

Blaikie, P. (1995) 'Understanding environmental issues', in S. Morse and M. Stocking (eds) *People and Environment*, London: UCL Press.

Blaikie, P. and Brookfield, H. (1987) *Land Degradation and Society*, London: Routledge.

Blanco, J. and Benjamin, M. (1994) *Talking About Revolution*, Melbourne: Ocean Press.

Blaug, M. (1968) *Economic Theory in Retrospect*, London: Heinemann.

Bleaney, M. (1985) *The Rise and Fall of Keynesian Economics*, London: Macmillan.

Block, W. (1990) 'Environmental problems, private property rights solutions', in W. Block (ed.) *Economics and the Environment: A Reconciliation*, Vancouver: Fraser Institute.

Blowers, A. (1984) *Something in the Air: Corporate Power and the Environment*, London: Harper & Row.

Boden, M. (1992) *The Creative Mind*, London: Abacus.

Boggs, C. (1976) *Gramsci's Marxism*, London: Pluto Press.

Bohm, D. and Hiley, B.J. (1975) 'On the intuitive understanding of non-locality as implied by quantum theory', *Foundations of Physics* No. 5.

Bookchin, M. (1990) *The Philosophy of Social Ecology*, Montreal: Black Rose Books.

Bookchin, M. (1997) 'Freedom of the city', *Red Pepper*, December.

Booth, D. (1985) 'Marxism and development sociology: interpreting the impasse', *World Development* Vol. 13.

Booth, D. (1994) 'How far beyond the impasse: a provisional summing up', in D. Booth *Rethinking Social Development: Theory, Research and Practice*, London: Longman.

Bottomore, T. (1985) *Theories of Modern Capitalism*, London: George Allen & Unwin.

Boulding, K.E. (1989) *Three Faces of Power*, Newbury Park, CA: Sage.

Boulding, K.E. (1991) 'What do we want to sustain? Environmentalism and human evaluations', in R. Costanza (ed.) *Ecological Economics: The Science and Management of Sustainability*, New York: Columbia University Press.

Bowers, J. (1997) *Sustainability and Environmental Economics: An Alternative Text*, London: Addison Wesley Longman.

Bremner, B. (1998) 'What to do about Asia', *Business Week*, 26 January.

Brett, E. (1985) *The World Economy Since the War*, London: Macmillan.

Broad, R. (1988) *Unequal Alliance: The World Bank, the International Monetary Fund, and the Philippines*, Berkeley: University of California Press.

Brogan, P. (1992) *World Conflicts: Why and Where They are Happening*, London: Bloomsbury.

Brohman, J. (1996) *Popular Development: Rethinking the Theory and Practice of Development*, London: Blackwell.

Bromley, D. (ed.) (1992) *Making the Commons Work: Theory Practice and Policy*, San Francisco: Institute of Contemporary Studies Press.

Bromley, D. and Cernea, M. (1989) 'The management of common property resources: some conceptual and operational fallacies', World Bank Discussion Paper No. 57, Washington: World Bank.

Brown, M.B. (1974) *The Economics of Imperialism*, Harmondsworth: Penguin.

Brunner, C. (1977) *Cuban Sugar Policy from 1963 to 1970*, Pittsburgh: Pittsburgh University Press.

Buchanan, J.M. and Vanberg, V.J. (1990) 'The market as a creative process', unpublished manuscript.

Burbridge, J. (ed.) (1997) *Beyond Prince and Merchant: Citizen Participation and the Rise of Civil Society*, New York: Pact.

Burgess, R. (1978) 'The concept of nature in geography and Marxism', *Antipode* 10(2).

Burkitt, B. (1984) *Radical Political Economy*, Brighton: Wheatsheaf.

Buttel, F. (1992) 'Environmentalism', *Rural Sociology* 57(1).

Buzan, T. (1974) *Use Your Head*, London: Ariel Books.

Buzan, T. (1995) *The Mind Map Book*, London: BBC Books.

Bygraves, M. (1998) 'From Wall Street to High Street', *Guardian*, 25 July.

Cabral, A. (1979) *Unity and Struggle*, New York: Monthly Review Press.

Callinicos, A. (1989) *Against Postmodernism*, Cambridge: Polity Press.

Callinicos, A. (1991) *The Revenge of History*, Cambridge: Polity Press.

Cameron, J. (1984) 'Theory and response to growing unemployment among young people in Fiji', in R. Fiddy (ed.) *Sixteen Years to Life: National Strategies for Youth Unemployment*, Brighton: Falmer Press.

Capra, F. (1982) *The Turning Point*, London: Flamingo.

Cardoso, O.J. (1997) 'A cheese for nobody', in P. Bush (ed.) *The Voice of The Turtle: An Anthology of Cuban Stories*, London: Quartet.

Carson, R. (1962) *Silent Spring*, New York: Fawcet Crest Books.

Carvalho, F. (1992) *Mr Keynes and the Post Keynesians*, Aldershot: Edward Elgar.

Cassen, R. (1994) 'Structural adjustment in sub-Saharan Africa', in W. van der Geest (ed.) *Negotiating Structural Adjustment in Africa*, London: Currey-Heinemann.

Castells, M. (1980) *The Economic Crisis and American Society*, Princeton: Princeton University Press.

Castro, F. (1972) *La Revolución Cubana 1953-1962*, Mexico City: Editorial ERA.

Castro, F. (1987) 'Discurso Pronunciado en la Clausulá de V. Congresso de la Unión de Jovenes Communistas, La Habana 17 Abril de 1987', *Cuba Socialista* No. 17.

Castro, F. (1991) *Presente y Futuro*, interview with the Mexican magazine *Siempre!* La Habana: Oficina de Publicaciones de Consejo de Estado.

Caufield, C. (1997) *Masters of Illusion: The World Bank and the Poverty of Nations*, London: Macmillan.

Chamberlain, E.H. (1933) *The Theory of Monopolistic Competition*, Cambridge, MA: Harvard University Press.

Chambers, R. (1997) *Whose Reality Counts? – Putting the Last First*, London: Intermediate Technology Publications.

Chase, A. (1980) *The Legacy of Malthus: The Social Cost of the New Scientific Racism*, Urbana, IL: University of Illinois Press.

Chenery, H. (1975) 'A structuralist approach to development policy', *American Economic Review* 65(2).

Chew, S.C. and Denemark, R.A. (eds) (1996) *The Underdevelopment of Development*, London: Sage.

Chomsky, N. (1996) *World Orders Old and New*, London: Pluto Press.

Clark, M.E. (1991) 'Rethinking ecological and economic education: a gestalt shift', in R. Costanza (ed.) *Ecological Economics: The Science and Management of Sustainabiity*, New York: Columbia University Press.

Cleaver, H. (1979) *Reading Capital Politically*, Brighton: Harvester.

Coase, R. (1960) 'The problem of social cost', *Journal of Law and Economics* Vol. III.

Coates, D. (1994) *The Question of UK Decline: The Economy, State and Society*, London: Harvester Wheatsheaf.

Coates, D. and Hillard, J. (eds) (1985) *UK Economic Decline: Key Texts*, London: Harvester Wheatsheaf.

Coates, D. and Hillard, J. (eds) (1986) *The Economic Decline of Modern Britain: The Debate Between Left and Right*, Brighton: Wheatsheaf.

Coates, K. (1972) 'Socialism and the environment', in K. Coates (ed.) *Socialism and the Environment*, Nottingham: Spokesman.

Cohen, J. and Stewart, I. (1994) *The Collapse of Chaos*, London: Penguin.

Colander, D. (1988) 'The evolution of Keynesian economics: from Keynesian to New Classical to New Keynesian', in O.F. Harmouda and J.N. Smithin (eds) *Keynes and Public Policy After Fifty Years*, Aldershot: Edward Elgar.

Colclough, C. (1982) 'Lessons from the development debate for Western economic policy', *Foreign Affairs* Vol. 58.

Cole, G.D.H. and Postgate, R. (1961) *The Common People*, London: Methuen.

Cole, K. (1983) 'Economic reality and development studies', *Journal Economi Malaysia* No. 7.

Cole, K. (1987) 'Reality, theory and the role of the agricultural sector in economic development', *Economic Analysis and Policy* 17(1).

Cole, K. (1993) *The Intellectual Parameters of the Real World*, Development Studies Discussion Paper No. 234, University of East Anglia, Norwich.

Cole, K. (1994) 'Ideologies of sustainable development', in K. Cole (ed.) *Sustainable Development for a Democratic South Africa*, London: Earthscan.

Cole, K. (1995) *Understanding Economics*, London: Pluto Press.

Cole, K. (1998) *Cuba: From Revolution to Development*, London: Pinter.

Cole, K., Cameron, J. and Edwards, C. (1991) *Why Economists Disagree*, London: Longman.

Cole, K. and Yaxley, I. (1991) *The Dialectics of Socialism*, Discussion Paper No. 220, School of Development Studies, University of East Anglia, Norwich.

Commoner, B. (1971) *The Closing Circle*, New York: Knopf.

Commoner, B. (1979) *The Politics of Energy*, New York: Knopf.

Commoner, B. (1990) *Making Peace with the Planet*, London: Gollancz.

Corbridge, S. (1986) *Capitalist World Development*, London: Macmillan.

Costanza, R., Daly, H.E. and Bartholomew, J.A. (1991) 'Goals, agenda, and policy recommendations for ecological economics', in R. Costanza (ed.) *Ecological Economics: The Science and Management of Sustainabiity*, New York: Columbia University Press.

Cowen, T. (1997) *Risk and Business Cycles: New and Old Austrian Perspectives*, London: Routledge.

Cox, A., Furlong, P. and Page, E. (1985) *Power in Capitalist Society*, Brighton: Wheatsheaf.

Cox, R. (1981) 'Social forces, states and world orders: beyond international relations theory', *Journal of International Studies: Millennium* 10(2).

Cox, R. (1993) 'Structural issues of global governance: implications for Europe', in S. Gill (ed.) *Gramsci, Historical Materialism and International Relations*, Cambridge: Cambridge University Press.

Craven, J. (1984) *Introduction to Economics*, Oxford: Blackwell.

Crook, C. (1991) 'The IMF and the World Bank', *The Economist*, 12 October.

Cronon, W. (1983) *Changes in the Land: Indians, Colonists, and the Ecology of New England*, New York: Hill and Wang.

Crook, S., Pakulski, J. and Waters, M. (1992) *Post-Modernization: Change in Advanced Societies*, London: Sage.

Cruz, W., Munasinghe, M. and Warford, J. (1997) *The Greening of Economic Policy Reform*, Vol. II: *Case Studies*, Washington: World Bank.

Daley-Harris, S. (1997) *The Microcredit Summit Report*, Washington: Results Educational Fund.

Daly, H. and Cobb, J. (1990) *For the Common Good*, London: Greenprint.

David, W. (1986) *Conflicting Paradigms in the Economics of Developing Nations*, New York: Praeger.

de Janvry, A. and Garcia, R. (1988) 'Rural poverty and environmental degradation in Latin America', paper presented to the IFAD Conference on Smallholders and Sustainable Development, Rome.

de Janvry, A., Sadoulet, E. and Thornebecke, E. (1991) 'States, markets and civil institutions', paper presented to the ILO Conference on States, Markets and Civil Society, Ithaca.

Denton, D. (1993) *The Pinnacle of Life*, London: Allen & Unwin.

Department of the Environment (1988) *Environmental Assessment*, London: HMSO.

Department of the Environment (1990) *The Common Inheritance*, London: HMSO.

Department of the Environment (1991) *Policy Appraisal and the Environment: A Guide for Government Departments*, London: HMSO.

Department of the Environment (1992) *The UK Environment*, London: HMSO.

Desai, M. (1974) *Marxian Economic Theory*, London: Gray-Mills.

de Santamaria, A. (1992) 'Economic science and political democracy', in P. Ekins and M. Max-Neef (eds) *Real-Life Economics*, London: Routledge.

Devall, B. (1988) *Simple in Means, Rich in Ends: Practicing Deep Ecology*, Salt Lake City: Peregrine Smith.

Devall, B. and Sessions, E. (1985) *Deep Ecology*, New York: Oxford University Press.

Diesing, P. (1982) *Science and Ideology in the Policy Sciences*, New York: Aldine.

Ditch, J. (1991) 'The undeserving poor: unemployed people then and now', in M. Loney, R. Bocock, J. Clarke, A. Cochrane, P. Graham and M. Wilson (eds) *The State or the Market*, London: Sage.

Dolan, E. and Lindsey, D. (1988) *Economics*, Chicago: Dreydon Press.

Douglas, M. (1980) 'Environment at risk', in J. Dowie and P. Lefrere (eds) *Risk and Change*, Milton Keynes: Open University Press.

Dragsted, A. (ed.) (1976) *Value: Studies by Marx*, London: New Park.

Drucker, P. (1954) *The Practice of Management*, New York: Harper & Row.

Dryzek, J. and Lester, J. (1989) 'Alternative views of the environmentalist problematic', in J. Lester (ed.) *Environmental Politics and Policy*, London: Duke University Press.

Duncan, G. (1983) 'Political theory and human nature', in I. Forbes and S. Smith (eds) *Politics and Human Nature*, London: Frances Pinter.

Echikson, W. (1990) *Lighting the Night*, London: Pan.

Ecologist (1974) *A Blueprint for Survival*, New York: Signet Books.

Edwards, S. and Van Wijnbergen, S. (1989) 'Disequilibrium and structural adjustment', in H. Chenery and T. Srinivasan (eds) *Handbook of Development Economics*, Amsterdam: North Holland.

Eichner, A.S. (ed.) (1987) *The Macrodynamics of Advanced Market Economies*, London: Sharpe.

Elliott, C. (1989) *Sword and Spirit: Christianity in a Divided World*, BBC, London.

Elshtain, J. (1985) 'Reflections on war and political discourse: realism, just war and feminism in a nuclear age', *Political Theory* 13(1).

Emmanuel, A. (1971) *Unequal Exchange: A Study of the Imperialism of Trade*, London: New Left Books.

Engels, F. (1940) *Dialectics of Nature*, New York: International Publishers.

Engels, F. (1973) *The Condition of the Working Class in England*, Moscow: Progress Publishers.

Engels, F. (1975) 'Extracts from *Anti-Duhring*', in K. Marx and F. Engels *Marx-Engels: On Religion*, Moscow: Progress Publishers.

Engels, F. (1977) *Principles of Communism*, Peking: Foreign Languages Press.

Environmental Protection Agency (1993) *Sourcebook for Environmental Assessment (EA)*, Washington: EPA.

Erlich, P. (1968) *The Population Bomb*, New York: Ballantine.

Erlich, P. and Erlich, A. (1990) *The Population Explosion*, New York: Simon & Schuster.

Espring-Anderson, G. (1990) *The Three Worlds of Welfare Capitalism*, Cambridge: Polity Press.

Esty, D. (1994) *Greening the GATT*, Washington: Institute of International Economics.

Evans, D. (1989) *Comparative Advantage and Growth: Trade and Development in Theory and Practice*, London: Harvester/Wheatsheaf.

Evenson, D. (1994) *Revolution in the Balance: Law and Society in Contemporary Cuba*, Boulder, CO: Westview Press.

Faber, D. (ed.) (1998) *The Struggle for Economic Democracy*, New York: Guilford.

Faber, M., Manstetten, R. and Proops, J. (1996) *Ecological Economics: Concepts and Methods*, Cheltenham: Elgar.

Farrington, J. and Bebbington, A. (1993) *Reluctant Partners?*, London: Routledge.

Fermont, C. (1998) 'Suharto's last stand', *Socialist Review* No. 218.

Feshbach, M. and Friendly, A. (1992) *Ecocide in the USSR*, New York: Basic Books.

Field, B. (1994) *Environmental Economics: An Introduction*, New York: McGraw-Hill.

Fine, R. (1990) *Beyond Apartheid*, London: Pluto Press.

Fischer, D.H. (1996) *The Great Wave: Price Revolutions and the Rhythm of History*, New York: Oxford University Press.

Fitton, L.J, Choe, J., Regan, R., Stolberg, R. and Butler, B. (eds) (1994) *Environmental Justice: Annotated Bibliography*, Washington: Center for Policy Alternatives.

Fitzgerald, F. (1994) *The Cuban Revolution in Crisis*, New York: Monthly Review Press.

Foot, P. (1998) 'Rich get richer', *Guardian*, 5 May: 14.

Foracs, D. (ed.) (1988) *A Gramsci Reader*, London: Lawrence & Wishart.

Foucoult, M. (1991) *Remarks on Marx*, New York: Semiotext.

Fowler, A. (1997) *Striking a Balance*, London: Earthscan.

Fox, W. (1990) *Towards a Transpersonal Ecology: Developing New Foundations for Environmentalism*, Boston: Shambhala.

Frank, A.G. (1969) *Capitalism and Underdevelopment in Latin America*, New York: Monthly Review Press.

Frank, A.G. (1972) *Lumpen-Bourgeosie – Lumpen-Development*, New York: Monthly Review Press.

Frank, A.G. (1996) 'The underdevelopment of development', in S.C. Chew and R.A. Denemark (eds) *The Underdevelopment of Development*, London: Sage.

Freire, P. (1975) *Pedagogy of the Oppressed*, London: Penguin.

Freire, P. (1993) *Pedagogy of the City*, New York: Continuum.

Friedman, J. (1979) *The Good Society*, Cambridge, MA: MIT Press.

Friedman, J. (1992) *Empowerment: The Politics of Alternative Development*, Oxford: Blackwell.

Friedman, M. (1953) *Essays in Positive Economics*, Chicago: University of Chicago Press.

Friedman, M. (1962) *Capitalism and Freedom*, Chicago: University of Chicago Press.

Friedman, M. (1968) 'The role of monetary theory', *American Economic Review* Vol. 58.

Friedman, M. (1974) *Monetary Correction*, London: Institute of Economic Affairs.

Friedman, M. (1976) 'The line we dare not cross', *Encounter*, November.

Friedman, M. and Friedman, R. (1979) *Free to Choose*, Harmondsworth: Penguin.

Friedmann, L. (1987) *Planning in the Public Domain*, Princeton: Princeton University Press.

Frisch, R. (1950) 'Alfred Marshall's theory of value', in H. Townsend (ed.) *Price Theory*, Harmondsworth: Penguin.

Fukuyama, F. (1992) *The End of History and the Last Man*, London: Hamish Hamilton.

Fyson, N. (1984) *The Development Puzzle*, London: Hodder & Stoughton.

Gadzey, A.T. (1994) *The Political Economy of Power: Hegemony and Economic Liberalism*, London: Macmillan.

Galbraith, J.K. and Salinger, N. (1981) *Almost Everyone's Guide to Economics*, Harmondsworth: Penguin.

Gare, A. (1995) *Post-Modernism and the Environmental Crisis*, London: Routledge.

Gasper, P. (1998) 'Bookwatch: Marxism and science', *International Socialism* No. 79.

Ghai, D. and Vivian, J.M. (eds) (1992) *Grassroots Environmental Action*, London: Routledge.

Gibbon, P. (1992) 'The World Bank and African poverty', *Journal of Modern African Studies* 30(2).

Gill, S. (ed.) (1993) *Gramsci, Historical Materialism and International Relations*, Cambridge: Cambridge University Press.

Gill, S. and Law, D. (1989) *The Global Political Economy: Perspectives, Problems and Policies*, London: Harvester.

Gillis, M., Perkins, D., Roemer, M. and Snodgrass, D. (1983) *Economics of Development*, New York: Norton.

Gilpin, R. (1975) *US Power and the Multinational Corporation*, New York: Basic Books.

Giroux, H. (1981) *Ideology, Culture and the Process of Schooling*, Philadelphia: Temple University Press.

Girvan, N. (1973) The development of dependency economics in the Caribbean and Latin America: a review and comparison', *Social and Economic Studies* Vol. 22.

Glass, J. and Johnson, W. (1989) *Economics: Progression, Stagnation or Degeneration*, London: Harvester.

Goetz, A. and O'Brien, D. (1995) 'Governing for the common wealth – the World Bank approach to poverty and governance', *IDS. Bulletin* 26(2).

Goldstein, J. (1988) *Long Cycles: Prosperity and War in the Modern Age*, New Haven: Yale University Press.

Goldstone, J. (1986) *Revolutions: Theoretical, Comparative and Historical Studies*, Orlando: Harcourt Brace Jovanovich.

Gonzalez M. (1987) 'Chile 1972–73: The workers united', in C. Barker (ed.) *Revolutionary Rehearsals*, London: Bookmarks.

Goodin, R. (1992) *Green Political Theory*, Cambridge: Polity Press.

Gore, A. (1992) *Earth in the Balance: Forging a New Common Purpose*, London: Earthscan.

Gough, I. (1979) *The Political Economy of the Welfare State*, London: Macmillan.

Gould, C. (1990) *Rethinking Democracy: Freedom and Social Co-operation in Politics, Economy and Society*, Cambridge: Cambridge University Press.

Goulet, D. (1971) 'Development or liberation?', in C. Wilber and K. Jameson (eds) (1992) *The Political Economy of Development and Underdevelopment*, New York: McGraw-Hill.

GPO (1980) *Global 2000 Report to the President*, Washington: Government Printing Office.

Gray, A. and Thompson, A.E. (1980) *The Development of Economic Doctrine*, London: Longman.

Green, D. (1995) *Silent Revolution*, London: Latin America Bureau.

Griffiths, W. (1994) 'Appropriate economic theory for the Caribbean', in H. Watson (ed.) *The Caribbean in the Global Political Economy*, Boulder, CO: Lynne Rienner.

Gross, P.R. and Levitt, N. (1998) *Higher Superstition: The Academic Left and Its Quarrels With Science*, Baltimore: Johns Hopkins Press.

Group of Green Economists (1992) *Ecological Economics*, London: Zed Books.

Grove, R. (1993) *Green Imperialism*, Cambridge: Cambridge University Press.

Gyoung-Hee, S. (1998) 'The crisis and worker's movement in South Korea', *International Socialism* No. 78.

Habermas, J. (1987) *Knowledge and Human Interests*, Cambridge: Polity Press.

Hahn, F. (1994) 'An intellectual retrospect', *Banca Nazionale del Lavora Quarterly Review*, pp 245-58.

Haila, Y. and Levins, R. (1992) *Humanity and Nature: Ecology, Science and Society*, London: Pluto Press.

Hall, R.E. and Taylor, J.B. (1991) *Macroeconomics*, New York: Norton.

Harcourt, G.L. (1969) 'Some Cambridge controversies in the theory of capital', *Journal of Economic Literature* Vol. 7.

Harcourt, G.L. (1972) *Some Cambridge Controversies in the Theory of Capital*, Cambridge: Cambridge University Press.

Harcourt, G.L. (1976) 'The Cambridge controversies: old ways and new horizons', in C. Sardoni (1992) *On Political Economists and Modern Political Economy: Selected Essays of G.L. Harcourt*, London: Routledge.

Harcourt, G.L. (1994) 'Capital theory controversies', in P. Arestis and M. Sawyer (eds) *The Elgar Companion to Radical Political Economy*, Aldershot: Edward Elgar.

Hardin, G. (1968) 'The tragedy of the commons: the population problem has no technical solution; it requires a fundamental extension in morality', *Science* No. 162.

Hardwick, P., Khan, B. and Langmead, J. (1994) *An Introduction to Modern Economics*, London: Longman.

Harman, C. (1988) *The Fire Last Time, 1968 and After*, London: Bookmarks.

Harman, C. (1997) 'Postmortem', *International Socialism*, November.

Harrigan, F. and McGregor, P. (1991) 'The macroeconomics of the Chicago School', in P. Mair and A. Miller (eds) *A Modern Guide to Economic Thought*, Aldershot: Edward Elgar.

Harris, L. (1985) 'The UK economy at the crossroads', in J. Allen and D. Massey (eds) *The Economy in Question*, London: Sage.

Harris, N. (1983) *Of Bread and Guns*, Harmondsworth: Penguin.

Harris N. (1998) 'Endangered tigers', *Red Pepper*, January.

Harriss, J. (ed.) (1982) *Rural Development*, London: Hutchinson.

Harriss, J., Hunter, J. and Lewis, C. (eds) (1995) *The New Institutional Economics and Third World Development*, London: Routledge.

Hart, P. (1988) *Youth Unemployment in Great Britain*, Cambridge: Cambridge University Press.

Harvey, D. (1982) *The Limits to Capital*, Oxford: Blackwell.

Harvey, D. (1993) 'The nature of the environment: the dialectics of social and environmental change', in R. Miliband and L. Panitch (eds) *Socialist Register 1993: Real Problems – False Solutions*, London: Merlin Press.

Harvey, D. (1996) *Justice, Nature and the Geography of Difference*, Oxford: Blackwell.

Hassan, I. (ed.) (1987) *The Postmodern Turn: Essays in Postmodern Theory and Culture*, Charleston: Ohio State University Press.

Hausman, D.M. (1988) 'An appraisal of Popperian methodology', in N. De Marchi (ed.) *The Popperian Legacy in Economics*, Cambridge: Cambridge University Press.

Haverman, R. (1973) 'Private power and federal policy', in R. Haverman and R. Hemrin (eds) *The Political Economy of Federal Policy*, New York: Harper & Row.

Hayek, F. (1940) *Individualism and Economic Order*, London: Routledge Kegan Paul.

Hayek, F. (1942) 'Scientism and the study of society', *Economica* Vol. VIII, and in F. Hayek (1944) *The Counter Revolution in Science*, Indianapolis: Liberty Press.

Hayek, F. (1978) *New Studies in Philosophy, Politics, Economics and the History of Ideas*, Chicago: University of Chicago Press.

Hayek, F. (1979) *The Counter-Revolution of Science*, Indianapolis: Liberty Press.

Heater, D. (1980) *World Studies*, London: Harrap.

Heilbroner, R. (1992) *The Worldly Philosophers*, New York: Touchstone.

Heilbroner, R. and Ford, A.M. (1971) *Is Economics Relevant: A Reader in Political Economics*, Pacific Palisades: Goodyear.

Heilbroner, R. and Milberg, W. (1995) *The Crisis of Vision in Modern Economic Thought*, Cambridge: Cambridge University Press.

Helburn, S. (1986) 'Economics and economics education: the selective use of discipline structures in economic curricula', in S. Hodkinson and D. Whitehead (eds) *Economics Education: Research and Development Issues*, London: Longman.

Held, D. (1993) *Prospects for Democracy*, Cambridge: Polity Press.

Helm, D. and Pearce, D. (1990) 'Assessment: economic policy towards the environment', *Oxford Review of Economic Policy* 6(1).

Helm, D. and Pearce, D. (1991) 'Economic policy towards the environment', in D. Helm (ed.) *Economic Policy Towards the Environment*, Oxford: Blackwell.

Heredia, F. (1993) 'Cuban Socialism: prospects and challenges', in Centro de Estudios Sobre America *The Cuban Revolution in the 1990s*, Boulder, CO: Westview Press.

Hermassi, E. (1978) 'Changing patterns of research in the Third World', *Annual Review of Sociology* No. 4.

Hernández, H. and Nikolenkov, V. (1985) 'El mechanismo economico de social-ismo', *Economia y Desarrollo* No. 68.

Hewitt de Aleantara, C. (1998) 'Uses and abuses of the concept of governance', *Social Science Journal* No. 155.

Hicks, J.R. (1935) *Value and Capital*, Cambridge: Cambridge University Press.

Hippler, J. (ed.) (1995) *The Democratisation of Disempowerment*, London: Pluto Press.

Hirschman, A. (1981) *Essays in Trespassing – Economics to Politics and Beyond*, Cambridge: Cambridge University Press.

Hirst, P. (1990) *Associative Democracy: New Forms of Economic and Social Governance*, Cambridge: Polity Press.

Hirst, P. and Thompson, G. (1996) *Globalization in Question: The International Economy and the Possibilities of Governance*, Cambridge: Polity Press.

HMSO (1972) *Pollution: Nuisance or Nemesis?*, London: HMSO.

Hoare, Q. and Nowell Smith, G. (eds) (1971) *Selections from the Prison Notebooks of Antonio Gramsci*, London: Lawrence & Wishart.

Hodgson, D. (1993) *The Mind Matters: Consciousness and Choice in a Quantum World*, Oxford: Clarendon Press.

Hodgson, G. (1997) 'The fate of the Cambridge capital controversy', in P. Arestis, G. Palma and M. Sawyer (eds) *Capital Controversy, Post-Keynesian Economics and the History of Economics*, London: Routledge.

Hodgson, G. (1998) 'The approach of institutional economics', *Journal of Economic Literature* Vol. XXXVI.

Hoggart, R. (1998) *The Tyranny of Relativism: Culture and Politics in Contemporary English Society*, New Brunswick: Transaction Books.

Holling, C.S. (1979) 'Myths of ecological stability', in G. Smart and W. Stanbury (eds) *Studies in Crisis Management*, Montreal: Butterworth.

Holling, C.S. (1986) 'The resilience of terrestrial ecosystems', in W. Clark and R. Munn (eds) *Sustainable Development and the Biosphere*, Cambridge: Cambridge University Press.

Holton, R. (1992) *Economy and Society*, London: Routledge.

Honeywell, M. (ed.) (1983) *The Poverty Brokers: The IMF and Latin America*, London: Latin America Bureau.

Hoogvelt, A. (1997) *Globalization and the Postcolonial World*, London: Macmillan.

Hope, A. and Trimmel, S. (1995) *Training for Transformation* (5 volumes) Gweru Zimbabwe: Mambo Press.

Huckle, J. (ed.) (1983) *Geographical Education: Reflection and Action*, Oxford: Oxford University Press.

Hughes, J., Martin, P. and Sharrock, W. (1995) *Understanding Classical Sociology*, London: Sage.

Hulme, D. and Moseley, P. (1996a) *Finance Against Poverty* Vol. 1, London: Routledge.

Hulme, D. and Moseley, P. (1996b) *Finance Against Poverty* Vol. 2, London: Routledge.

Hunt, D. (1989) *Economic Theories of Development*, London: Harvester-Wheatsheaf.

Huntingdon, S. (1984) 'Will countries become more democratic', *Political Science Quarterly*, 99.

Hutchinson, T. (1981) *The Politics and Philosophy of Economics*, London: Blackwell.

Hutton, J. (1979) *The Mystery of Wealth*, Cheltenham: Stanley Thornes.

Hyden, G. (1983) *No Short Cut To Progress*, London: Heinemann.

Iliffe, J. (1983) *The Emergence of African Capitalism*, London: Longman.

IMF (1974) 'Decision 4377–(74/114) on Extended Fund Provision', in IMF (1990) *Selected Decisions*, Washington: International Monetary Fund.

Ionescu, G. and Gellner, E. (1969) *Populism: Its Meanings and National Character-istics*, London: Weidenfeld & Nicolson.

Irwin, A. (1995) *Citizen Science: A Study of People, Expertise and Sustainable Development*, London: Pluto Press.

Jagger, A. (1983) *Feminist Politics and Human Nature*, Brighton: Harvester.

Jain, P. (1996) 'Managing credit for the rural poor: lessons from the Grameen Bank', *World Development* Vol. 24.

Jansen, K. (ed.) (1986) *Monetarism, Economic Crisis and the Third World*, London: Frank Cass.

Jevons, W.S. (1970) *The Theory of Political Economy*, Harmondsworth: Penguin.

Johnson, H. and Bernstein, H. (1982) *Third World Lives of Struggle*, London: Heinemann.

Johnson, S. and Rogaly, B. (1997) *Microfinance and Poverty Reduction*, Oxford: OXFAM.

Johnston, B. and Clarke, D. (1991) *Redesigning Rural Development*, Baltimore: Johns Hopkins University.

Johnston, R. (1983) *Philosophy and Human Geography*, London: Edward Arnold.

Johnston, R. (1996) *Nature, State and Economy: The Political Economy of the Environment*, Chichester: Wiley.

Jones, S. (1994) *The Language of the Genes*, London: Flamingo.

Jones, S. (1996) *In the Blood: God, Genes and Destiny*, London: HarperCollins.

Kay, G. (1975) *Development and Underdevelopment: A Marxist Analysis*, London: Macmillan.

Keat, R. and Urry, J. (1982) *Social Theory as Science*, London: Routledge and Kegan Paul.

Keil, R., Bell, D.V.J, Penz, P. and Fawcett, L. (eds) (1998) *Political Ecology: Global and Local*, London: Routledge.

Keller, E.F. (1985) *Reflections on Gender and Science*, New Haven: Yale University Press.

Kellogg, P. and Whitney, S. (1998) *The MAI and Capitalist Crisis – A Marxist Analysis*, Toronto: International Socialists.

Kennedy, P. (1993) *Preparing for the Twenty-First Century*, London: Harper-Collins.

Kennedy, W.V. (1988) 'Environmental impact assessment in North America and Western Europe: what has worked where, how and why', *International Environment Reporter* 11(4).

Keynes, J.M. (1936) *The General Theory of Employment, Interest and Money*, London: Macmillan.

Khan, H. (1979) *World Economic Growth: 1979 and Beyond*, Boulder, CO: Westview Press.

Killick, T. (1984) *The Quest for Economic Stabilization*, London: Heinemann.

Kimber, C. (1998) 'Will giant firms rule the world', *Socialist Worker* 14 March.

Kindleberger, C. and Herrick, B. (1977) *Economic Development*, New York: McGraw-Hill.

Kirkpatrick, C. (1987) 'Trade policy and industrialization in LDCs', in N. Gemmel (ed.) *Surveys in Development Economics*, Oxford: Blackwell.

Kirzner, I.M. (1985) *Discovery and the Capitalist Process*, Chicago: University of Chicago Press.

Kirzner, I.M. (1992) *The Meaning of Market Process*, London: Routledge.

Kirzner, I.M. (1994a) *Austrian Economics: The Founding Era*, London: Pickering & Chatto.

Kirzner, I.M. (1994b) *Austrian Economics: The Interwar Period*, London: Pickering & Chatto.

Kirzner, I.M. (1994c) *Austrian Economics: The Age of Mises and Hayek*, London: Pickering & Chatto.

Kirzner, I.M. (1997) *How Markets Work: Disequilibrium, Entrepreneurship and Discovery*, London: Institute of Economic Affairs.

Kitching, G. (1982) *Development and Underdevelopment in Historical Perspective*, London: Methuen.

Kitromilides, Y. (1985) 'The formation of economic policy', in P. Arestis and T. Skouras (eds) *Post-Keynesian Economic Theory*, London: Wheatsheaf.

Kovel, J. (1995) 'Ecological Marxism and dialectics', *Capitalism Nature Socialism* 6(4).

Kristol, I. (1981) 'Rationalism in economics', in D. Bell and I. Kristol (eds) *The Crisis in Economic Theory*, New York: Basic Books.

Kuhn, T. (1970) *The Structure of Scientific Revolutions*, Chicago: University of Chicago Press.

Kuttner, R. (1985) 'The poverty of economics', *Atlantic Monthly*, February.

Labriola, A. (1904) *Essays in the Materialist Conception of History*, Chicago: Kerr.

Laclau, E. (1971) 'Feudalism and capitalism in Latin America', *New Left Review* No. 67.

Lage, C. (1996) 'The effects and results of the first half of the year are mainly towards the solution of the economy's fundamental problems', *Granma International*, 7 August.

Lal, D. (1976) 'Distribution and development: a review article', *World Development* 4(9).

Lal, D. (1983) *The Poverty of Development Economics*, London: Institute of Economic Affairs.

Lancaster, K. (1974) *Introduction to Modern Microeconomics*, Chicago: Rand McNally.

Lancaster, K. and Lipsey, R.G. (1956) 'The general theory of the second best', *Review of Economic Studies*, December.

Lappé, F.M. and Collins, J. (1977) *Food First: The Myth of Scarcity*, London: Souvenir Press.

Lappé, F.M. and Du Bois, P. (1994) *The Quickening of America*, New York: Simon & Schuster.

Lappé, F.M. and Schurman, R. (1989) *Taking Population Seriously*, London: Earthscan.

Laver, M. (1981) *The Politics of Private Desires*, Harmondsworth: Penguin.

Lavoie, M. (1992) *Foundations of Post-Keynesian Economic Analysis*, Aldershot: Edward Elgar.

Lawson, T. (1992) 'Abstraction, tendencies and stylized facts', in P. Ekins and M. Max-Neef (eds) *Real-Life Economics*, London: Routledge.

Lawson, T. (1997) *Economics and Reality*, London: Routledge.

Layder, D. (1994) *Understanding Social Theory*, London: Sage.

Leach, M. and Mearns, R. (1996a) *The Lie of the Land*, Oxford: Currey.

Leach, M. and Mearn, R. (1996b) 'Challenging received wisdom in Africa', in M. Leach and R. Mearn *The Lie of the Land*, Oxford: Currey.

Lebowitz, M.A. (1992) 'Capitalism: how many contradictions', *Capitalism Nature Socialism* 3(3).

Lee, D. and Newby, H. (1987) *The Problem of Sociology*, London: Hutchinson.

Leff, E. (1992) 'A second contradiction of capitalism', *Capitalism Nature Socialism* 3(4).

Lélé, S. (1991) 'Sustainable development: a critical review', *World Development* 19(6).

Lensink, R. (1996) *Structural Adjustment in Sub-Saharan Africa*, London: Longman.

Leopold, A. (1968) *A Sand Loving Almanac*, New York: Oxford University Press.

le Roux, P. (1990) 'The case for a social democratic compromise', in N. Nattrass and E. Ardington (eds) (1990) *The Political Economy of South Africa*, Cape Town: Oxford University Press.

LeShan, L. and Margeneau, H. (1982) *Einstein's Space and Van Gogh's Sky*, London: Collier.

Lester, J. (1989) *Environmental Politics and Policy*, London: Duke University Press.

Levacic, R. and Rebman, A. (1982) *Macroeconomics: An Introduction to Keynesian-Neoclassical Controversies*, London: Macmillan.

Levenstein, C. and Wooding, J. (eds) (1998) *Work, Health, and Environment: Old Problems, New Solutions*, New York: Guilford.

Levine, D.P. (1977) *Economic Studies: Contribution to a Critique of Economic Theory*, London: Routledge and Kegan Paul.

Levins, R. and Lewontin, R. (1985) *The Dialectical Biologist*, Cambridge, MA: Harvard University Press.

Levins, R. and Lewontin, R. (1994) 'Holism and reductionism in ecology', *Capitalism Nature Socialism* 5(4).

Lewis, W.A. (1954) 'Economic development with unlimited supplies of labour', *Manchester School* Vol. 22.

Lewontin, R.C. (1982) *Human Diversity*, New York: Scientific American.

Lewontin, R.C. (1991) *The Doctrine of DNA: Biology and Ideology* , Harmondsworth: Penguin.

Lewontin, R. and Levins, R. (1996) 'The end of natural history', *Capitalism Nature Socialism* 7(1).

Lewontin, R. and Levins, R. (1997a) 'Organism and environment', *Capitalism Nature Socialism* 8(2).

Lewontin, R. and Levins, R. (1997b) 'Chance and necessity', *Capitalism Nature Socialism* 8(1).

Lewontin, R. and Levins, R. (1997c) 'The biological and the social', *Capitalism Nature Socialism* 8(3).

Leys, C. (1996) *The Rise and Fall of Development Theory*, Nairobi: EAEP; London: James Currey.

Lincoln, Y. and Guba, E. (1985) *Naturalistic Inquiry*, Beverley Hills, CA: Russell Sage.

Lipsey, R. (1966) *An Introduction to Positive Economics*, London: Weidenfeld & Nicolson.

Lipton, M. (1977) *Why Poor People Stay Poor: A Study of Urban Bias in World Development*, London: Temple Smith.

Lisk, F. (1977) 'Conventional strategies and basic needs fulfilment', *International Labour Review* 15(2).

Little, R. and Smith, M. (1991) 'Introduction', in R. Little and M. Smith (eds) *Perspectives on World Politics*, London: Routledge.

Loney, M., Bocock, R., Clarke, J., Cochrane, A., Graham, P. and Wilson, M. (1991) *The State or the Market*, London: Sage.

Long, N. (1992a) 'From Paradigm Lost to Paradigm Regained?', in N. Long and A. Long (eds) *Battlefields of Knowledge*, London: Routledge.

Long, N. (1992b) 'Introduction', in N. Long and A. Long (eds) *Battlefields of Knowledge*, London: Routledge.

Love, J. (1991) 'The orthodox Keynesian school', in P. Mair and A. Miller (eds) *A Modern Guide to Economic Thought*, Aldershot: Edward Elgar.

Lovejoy, A.O. (1964) *The Great Chain of Being*, Cambridge, MA: Harvard University Press.

Lovelock, J. (1989) *The Ages of Gaia: A Biography of Our Living Earth*, Oxford: Oxford University Press.

Lovelock, J. (1991) *Gaia: The Practical Science of Planetary Medicine*, London: Gaia Books.

Lovelock, J. and Margulis, L. (1974) 'Atmospheric homeostasis by and for the biosphere: the Gaia hypothesis', *Tellus* No. 6.

Lukes, S. (1977) *Power: A Radical View*, London: Macmillan.

Lutz, M. (1992) 'Humanistic economics: history and basic principles', in P. Ekins and M. Max-Neef (eds) *Real-Life Economics*, London: Routledge.

MacArthur, B. (1998) *The Penguin Book of Twentieth Century Protest*, London: Viking.

Machel, S. (1975) *Samora Machel Speaks*, New York: Black Liberation Press.

Mackintosh, M. (1992) 'Introduction', in M. Wuyts, M. Mackintosh and T. Hewitt (eds) *Development Policy and Public Action*, Oxford: Oxford University Press.

Magnum, G., Magnum, S. and Phillips, P. (1987) 'The three (at least) worlds of economic theory', *Challenge*, March/April.

Malnes, R. (1995) *Valuing the Environment*, London: Manchester University Press.

Malthus, T. (1872) *Essay on the Principle of Population* 7th edn, London: Dent.

Mandel, E. (1977) *Long Waves of Capitalist Development: The Marxist Interpretation*, Cambridge: Cambridge University Press.

Mandel, E. (1978) *The Second Slump*, London: Verso.

Mandel, E. (1983) 'Economics', in D. McLellan (ed.) *Marx: The First 100 Years*, London: Fontana.

Mannathoko, C. (1992) 'Feminist theories and the study of gender issues in Southern Africa', in R. Meena (ed.) *Gender in Southern Africa: Conceptual and Theoretical Issues*, Harare: Sapes.

Markandya, A. and Richardson, J. (eds) (1992) *The Earthscan Reader in Environmental Economics*, London: Earthscan.

Marshall, A. (1947) *Principles of Economics*, London: Macmillan.

Marshall, M. (1987) *Long Waves in Regional Development*, London: Macmillan.

Martinez-Aller, J. (1990) *Ecological Economics: Energy, Environment, Society*, Oxford: Blackwell.

Marx, K. (1936) *The Poverty of Philosophy*, London: Lawrence & Wishart.

Marx, K. (1950) 18th Brumaire, in K. Marx and F. Engels *Collected Works* Vol. 1, London: Lawrence & Wishart.

Marx, K. (1970) *Critique of Hegel's 'Philosophy of Right'*, London: Lawrence & Wishart.

Marx, K. (1972a) *Capital* Vol. III, London: Lawrence & Wishart.

Marx, K. (1972b) *Critique of the Gotha Programme*, Peking: Foreign Languages Press.

Marx, K. (1973) *Grundisse*, Harmondsworth: Penguin.

Marx, K. (1974) *Capital* Vol. I, London: Lawrence & Wishart.

Mark, K. (1975a) *Wages, Price and Profit*, Peking: Foreign Languages Press.

Marx, K. (1975b) *Early Writings*, London: Penguin.

Marx, K. (1976a) *A Contribution to the Critique of Political Economy*, London: Lawrence & Wishart.

Marx, K. (1976b) *Preface and Introduction to a Contribution to the Critique of Political Economy*, Peking: Foreign Languages Press.

Marx, K. and Engels, F. (1952) *Manifesto of the Communist Party*, Moscow: Progress Publishers.

Marx, K. and Engels, F. (1970) *The German Ideology: Part One*, London: Lawrence & Wishart.

Marx, K. and Engels, F. (1975) *Marx-Engels: On Religion*, Moscow: Progress Publishers.

May, B. (1981) *The Calamity of the Third World*, London: Routledge and Kegan Paul.

McGrath, M. (1990) 'Economic growth, income distribution and social change', in N. Nattrass and E. Ardington (eds) *The Political Economy of South Africa*, Cape Town: Oxford University Press.

McKinley, R. and Little, R. (1986) *Global Problems and World Order*, London: Pinter.

McNally, D. (1993) *Against the Market*, London: Verso.

Meadows, D.H., Meadows, D.L., Randers, J. and Behrens, W. (1974a) *The Limits to Growth*, London: Pan.

Meadows, D.H., Meadows, D.L., Randers, J. and Behrens, W. (1974b) *Dynamics of Growth in a Finite World*, Cambridge, MA: Wright-Allen.

Meadows, D.H., Meadows, D.L. and Randers, J. (1992) *Beyond the Limits*, London: Earthscan.

Medin, T. (1990) *Cuba: The Shaping of Revolutionary Consciousness*, Boulder, CO: Reinner.

Meek, J. (1998) 'Russia's rouble panic may lead to a comeback for the Iron Curtain' *Observer* 16 August.

Meek, R.C. (1953) *Marx and Engels on Malthus*, London: Lawrence & Wishart.

Meier, G.M. (ed.) (1995) *Leading Issues in Economic Development*, Oxford: Oxford University Press.

Mellos, K. (1988) *Perspectives on Ecology*, London: Macmillan.

Menger, C. (1981) *Principles of Economics*, New York: New York University Press.

Merchant, C. (1980) *The Death of a Nation*, New York: Harper & Row.

Merchant, C. (1992) *Radical Ecology*, London: Routledge.

Midgely, M. (1978) *Of Beast and Man*, Brighton: Harvester.

Miles, M. (1995) 'Gender and global capitalism', in L. Sklair (ed.) *Capitalism and Development*, London: Routledge.

Miliband, R. and Panitch, L. (eds) (1993) *Socialist Register 1993: Real Problems – False Solutions*, London: Merlin Press.

Mises, L. von (1960) *Epistemological Problems of Economics*, Princeton NJ: Van Nostrand.

Mishan, E. (1971) *Cost–Benefit Analysis*, London: George Allen & Unwin.

Mishan, E. (1993) 'Economists versus the Greens: an exposition and critique', *Political Quarterly* 64(2).

Moggridge, D.E. (1976) *Keynes*, London: Fontana/Collins.

Mohun, S. (1989) 'Continuity and change in state economic intervention', in A. Cochrane and J. Anderson (eds) *Politics in Transition*, London: Sage.

Montgomery, R., Davis, R., Saxena, N.C. and Ashley, S. (1996) *Guidance Material for Improved Project Monitoring and Impact Review System in India*, Swansea: Centre for Development Studies.

Morrison, K. (1995) *Marx, Durkheim, Weber*, London: Sage.

Morton, A.L. (1965) *A People's History of England*, London: Lawrence & Wishart.

Mubende, K. (1986) *Namibia: The Broken Shield – Anatomy of Imperialism and Social Revolution*, Malme: Liber Forlag.

Mueller, D. (1991) *Choosing a Constitution in Eastern Europe: Lessons for Public Choice*, University of Maryland: mimeo.

Munasinghe, M., Cruz, W. and Warford, J. (1993) 'Are economy wide policies good for the environment', *Finance and Development* 30(3).

Muñiz, M. (ed.) (1993) *Elecciones en Cuba*, Melbourne: Ocean Press.

Murray, M.J. (1994) *The Revolution Deferred: The Painful Birth of Post-Apartheid South Africa*, London: Verso.

Naess, A. (1973) 'The shallow and the deep, long-range ecology movement: a summary', *Inquiry* No. 16.

Naess, A. (1989) *Ecology, Community and Lifestyle: Outline of an Ecosophy* (translated and edited by D. Rothenberg), Cambridge: Cambridge University Press.

Narda, M. (1991) 'Is modern science a Western patriarchal myth? A critique of populist orthodoxy', *South Asia Bulletin* Vol. XI.

Nash, F. (1989) *The Rights of Nature*, New York: Oxford University Press.

Nell, E.J. (1996) *Making Sense of a Changing Economy: Technology, Markets and Morals*, London: Routledge.

Norgaard, R. (1985) 'Environmental problems: an evolutionary critique and a plea for northpluralism', *Journal of Environmental Economics and Management* Vol. 12.

North, D. (1990) *Institutions, Institutional Change and Economic Performance*, Cambridge: Cambridge University Press.

North, D. (1995) 'The new institutional economics and Third World development', in J. Harriss, J. Hunter and C. Lewis (eds) *The New Institutional Economics and Third World Development*, London: Routledge.

Norris, C. (1992) *Uncritical Theory*, London: Lawrence & Wishart.

Nyerere, J. (1962) 'Ujamaa – the basis of African socialism', in J. Nyerere (1967) *Freedom and Unity: uhuru na umoja*, Oxford: Oxford University Press.

O'Brien, P. (1975) 'A critique of Latin American theories of dependency', in I. Oxall, T. Barnett and D. Booth (eds) *Beyond the Sociology of Development*, London: Routledge and Kegan Paul.

O'Connor, J. (1988) 'Capitalism, nature, socialism – a theoretical introduction', *Capitalism Nature Socialism* 1(1).

OECD (1989) *Economic Instruments for Environmental Control*, Paris: Organization for Economic Cooperation and Development.

Oliver, M. (1996) 'A sociology of disability or a disablist sociology?', in L. Barton (ed.) *Disability and Society*, London: Longman.

Ollman, G. (1976) *Alienation*, Cambridge: Cambridge University Press.

Ollman, G. (1993) *Dialectical Investigations*, London: Routledge.

Ormerod, P. (1994) *The Death of Economics*, London: Faber.

Ostrom, E. (1990) *Governing the Commons: The Evolution of Institutions for Collective Action*, Cambridge: Cambridge University Press.

OXFAM (1998) *For Richer For Poorer*, OXFAM Briefing Notes, Oxford: OXFAM.

Pacific Asia Resource Centre (1993) *The People vs Global Capital: The G7, TNCs,*

SAPs, and Human Rights, New York: Apex Press.

Parsons, H.L. (1977) *Marx and Engels on Ecology*, London: Greenwood.

Pateman, C. (1970) *Participation and Democratic Theory*, Cambridge: Cambridge University Press.

Pearce, D. (1983) *Cost–Benefit Analysis*, London: Macmillan.

Pearce, D. (ed.) (1993a) *Blueprint 2: Greening the World Economy*, London: Earthscan.

Pearce, D. (1993b) *Blueprint 3: Measuring Sustainable Development*, London: Earthscan.

Pearce, D. (1995) *Blueprint 4: Capturing Global Environmental Value*, London: Earthscan.

Pearce, D., Markandya, A. and Barbier, E. (1989) *Blueprint for a Green Economy*, London: Earthscan.

Pearce, D. and Turner, K. (1990) *Economics of Natural Resources and the Environment*, London: Harvester-Wheatsheaf.

Pearce, I. (1989) 'Putting perspectives in perspective', *Economics* XXV(1).

Pearce, J. (1986) *Promised Land*, London: Latin America Bureau.

Pearson, L. (1969) *Partners in Development*, New York: Praeger.

Peet, R. (1991) *Global Capitalism: Theories of Social Development*, London: Routledge.

Peet, R. and Watts, M. (eds) (1996a) *Liberation Ecologies: Environment, Development and Social Movements*, London: Routledge.

Peet, R. and Watts, M. (1996b) 'Liberation ecology: development, sustainability and environment in an age of market triumphalism', in R. Peet and M. Watts (eds) *Liberation Ecologies: Environment, Development and Social Movements*, London: Routledge.

Pepper, D. (1993) *Eco-Socialism: From Deep Ecology to Social Justice*, London: Routledge.

Pepper, D. (1996) *Modern Environmentalisn: An Introduction*, London: Routledge.

Peréz, L. (1982) *Cuba: Between Reform and Revolution*, Oxford: Oxford University Press.

Petras, J. (1978) *Critical Perspectives on Imperialism and Class in the Third World*, London: Monthly Review Press.

Petras, J. (1998) 'A Marxist critique of post-Marxism', *Link* No. 9.

Petrovic, G. (1983) 'Alienation', in T. Bottomore (ed.) *A Dictionary of Marxist Thought*, Oxford: Blackwell.

Pheby, J. (ed.) (1989) *New Directions in Post Keynesian Economics*, Aldershot: Edward Elgar.

Phillips, A.W. (1958) 'The relation between unemployment and the rate of change in money wage rates in the UK 1957–1986', *Economica* Vol. XXV.

Pilger, J. (1994) *Distant Voices*, London: Vintage.

Pilger, J. (1998) *Hidden Agendas*, London: Vintage.

Pilgrim, D. and Rogers, A. (1994) *A Sociology of Mental Health and Illness*, Buckingham: Open University Press.

Polak, J.J. (1996) *The World Bank and the International Monetary Fund: A Changing Relationship*, Washington: Brookings Institute.

Popper, K. (1960) *The Poverty of Historicism*, London: Routledge and Kegan Paul.

Portney, P. (1990) *Public Policies for Environmental Protection*, Washington: Resources for the Future.

Pourgerami, A. (1990) *Development and Democracy in the Third World*, London: Routledge.

Poya, M. (1987) 'Iran 1979: Long live the revolution. Long live Islam?', in C. Barker (ed.) *Revolutionary Rehearsals*, London: Bookmarks.

Prebisch, R. (1950) *The Economic Development of Latin America and the Principal Problems*, New York: Economic Commission for Latin America.

Prebisch, R. (1962) 'The economic development of Latin America and its principal problems', *Economic Bulletin of Latin America* VII(1).

Preston, P. (1982) *Theories of Development*, London: Routledge and Kegan Paul.

Putman, H. (1983) *Realism and Reason: Philosophical Papers*, Cambridge: Cambridge University Press.

Ramphele, M. (1991) *Restoring the Land*, London: Panos.

Ravaioli, C. (1995) *Economists and the Environment*, London: Zed Books.

Reason, P. and Rowan, J. (eds) (1981) *Human Inquiry*, New York: John Wiley.

Redclift, M. (1987) *Sustainable Development*, London: Routledge.

Redclift, M. and Benton, T. (1994) *Social Theory and the Global Environment*, London: Routledge.

Reder, M. W. (1982) 'Chicago economics: permanence and change', *Journal of Economic Literature* Vol. 20.

Ree, J. (1998) 'The storm is put back in the teacup', *Observer* 10 July.

Reed, D. (1992) *Structural Adjustment and the Environment*, London: Earthscan.

Reed, G. (1992) *Island in the Storm*, Melbourne: Ocean Press.

Rees, J. (1998) *The Algebra of Revolution*, London: Routledge.

Reich, R. (1998) 'Light blue touchpaper', *Observer* 15 March.

Reiss, E. (1997) *Marx: A Clear Guide*, London: Pluto Press.

Reynolds, H. Jnr (1993) 'Preface', in P. Freire *Pedagogy of the City*, New York: Continuum.

Reynolds, L. (1971) *The Three Worlds of Economics*, New Haven: Yale University Press.

Rich, B. (1994) *Mortgaging the Earth*, London: Earthscan.

Ritzer, G. (1975) *Sociology: A Multi-Paradigm Science*, Boston: Allyn & Bacon.

Rius (1984) *Nicaragua for Beginners*, London: Writers and Readers.

Rizvi, S.A.T. (1994) 'The microfoundations project in general equilibrium theory', *Cambridge Journal of Economics* Vol. 18.

Robbins, L. (1984) *An Essay on the Nature and Significance of Economic Science*, London: Macmillan.

Robinson, J. (1933) *The Economics of Imperfect Competition*, London: Macmillan.

Robinson, J. (1953) 'The production function and the theory of capital', in G.L. Harcourt and N.F. Laing (eds) (1971) *Capital and Growth*, Harmondsworth: Penguin.

Robinson, J. (1971) *Economic Heresies*, Harmondsworth: Penguin.

Robinson, J. and Williamson, F. (1983) 'Ideology and logic', in F. Vicarelli (ed.) *Keynes's Relevance Today*, London: Macmillan.

Robinson, P. (1987) 'Portugal 1974-5: popular power', in C. Barker (ed.) *Revolutionary Rehearsals*, London: Bookmarks.

Rodríguez, J.L. (1990) *Estrageia de Desarrollo Económico en Cuba*, La Habana: Editorial de Ciencias Sociales.

Roll, E. (1973) *A History of Economic Thought*, London: Faber.

Romano, R. and Leiman, K. (1970) *Views on Capitalism*, Beverley Hills: Glencoe Press.

Romer, D. (1993) 'The new Keynesian synthesis', *Journal of Economic Perspectives* 7(1).

Rose, S. (1973) *The Conscious Brain*, London: Weidenfeld & Nicolson.

Rose, S. (1984) 'Introduction', in S. Rose and L. Appignanesi (eds) *Science and Beyond*, Oxford: Blackwell.

Rose, S. (1997) *Lifelines: Biology, Freedom, Determinism*, London: Allen Lane.

Rose, S., Kamin, L.J. and Lewontin, R.C. (1984) *Not in Our Genes: Biology, Ideology and Human Nature*, Harmondsworth: Penguin.

Rostow, W.W. (1960) *The Stages of Economic Growth: An Anti-Communist Manifesto*, Cambridge: Cambridge University Press.

Routh, G. (1975) *The Origin of Economic Ideas*, London: Macmillan.

Rowthorn, B. (1974) 'Neo-classicism, neo-Ricardianism and Marxism', *New Left Review* No. 86.

Russett, B. and Starr, H. (1996) *World Politics: The Menu for Choice*, New York: Freeman.

Sand, P. (1995) 'Trusts for the Earth', in W. Lang (ed.) *Sustainable Development and International Law*, London: Graham and Trotman.

Saul, J. (1994) 'Globalism, socialism and democracy in the Southern African TRANSITION', in R. Miliband and L. Panitch (eds) *Socialist Register 1994*, London: Merlin.

Sayer, A. (1983) 'Notes on geography and the relation between people and nature', in T. Cannon, M. Forbes and J. Mackie (eds) *Society and Nature*, London: Union of Socialist Geographers.

Schelling, T.C. (ed.) (1983) *Incentives for Environmental Protection*, Cambridge, MA: MIT Press.

Schotter, A. (1990) *Free Market Economics*, London: Sage.

Schumacher, E.F. (1973) *Small is Beautiful*, London: Abacus.

Schuurman, F.J. (1993) 'Introduction: development theory and the 1990s', in F.J. Schuurman (ed.) *Beyond the Impasse*, London: Zed Books.

Seabrook, J. (1993) *Victims of Development*, London: Verso.

Seers, D. (1983) *The Political Economy of Nationalism*, Oxford: Oxford University Press.

Seidman, S. (1998) *Contested Knowledge: Social Theory in the Postmodern Era*, Oxford: Blackwell.

Seitz, J. (1988) *Politics of Development*, Oxford: Blackwell.

Seitz, J. (1995) *Global Issues: An Introduction*, Oxford: Blackwell.

Sender, J. and Smith, S. (1988) *The Development of Capitalism in Africa*, London: Methuen.

Shand, A. (1990) *Free Market Morality: The Political Economy of the Austrian School*, London: Routledge.

Shaw, T. and Gouldery, P. (1982) 'Alternative scenarios for Africa', in T. Shaw (ed.) *Alternative Futures for Africa*, Boulder, CO: Westview Press.

Shepherd, G. (1989) 'The reality of the commons: answering Hardin from Somalia', *Development Policy Review* No. 7.

Sherden, W.A. (1998) *The Fortune Sellers*, New York: Wiley.

Shivji, I. (1991) 'State and constitutionalism: an African debate on democracy', in I. Shivji (ed.) *Contradictory Class Perspectives in the Debate on Democracy*, Harare: Sapes.

Silver, B.L. (1998) *The Ascent of Science*, Oxford: Oxford University Press.

Simon, J. (1981) *The Ultimate Resource*, Princeton: Princeton University Press NJ.

Simon, J. and Khan, H. (1984) *The Resourceful Earth*, Oxford: Blackwell.

Simon, R. (1991) *Gramsci's Political Thought*, London: Lawrence & Wishart.

Simpson, J. (1990) *Dispatches from the Barricades*, London: Hutchinson.

Sloman, J. (1991) *Economics*, Hemel Hempstead: Harvester-Wheatsheaf.

Smith, A. (1970) *The Wealth of Nations*, Harmondsworth: Penguin.

Smith, D. (1987) *The Rise and Fall of Monetarism*, Harmondsworth: Penguin.

Smith, L.G. (1993) *Impact Assessment and Sustainable Resource Management*, London: Longman.

Smith, N. (1990) *Uneven Development: Nature, Capital and the Production of Space*, Oxford: Blackwell.

Smith, S. (1983) 'Introduction', in I. Forbes and S. Smith (eds) *Politics and Human Nature*, London: Frances Pinter.

Smith, S. and Toye, J. (eds) (1979) *Trade and Poor Economies*, London: Frank Cass.

So, A. (1990) *Social Change and Development*, London: Sage.

Sokal, A. and Bricmont, J. (1998) *Intellectual Impostures*, Profile Books, London.

Sowell, T. (1987) *A Conflict of Visions*, New York: William Morrison.

Sparks, C. (1998) 'The eye of the storm', *International Socialism* No. 78.

Sperry, R. (1980) 'Mind-brain interaction: mentalism yes, dualism no!', *Neuroscience* No. 5.

Sraffa, P. (1960) *Production of Commodities by Means of Commodities*, Cambridge: Cambridge University Press.

Srinivasan, T. (1989) 'Economic development – concepts and approaches: an introduction', in H. Chenery and T. Srinivasan (eds) *Handbook of Development Economics*, Amsterdam: North Holland.

Staniland, M. (1985) *What is Political Economy*, New Haven: Yale University Press.

Stankiewicz, W. (1992) *In Search of a Political Philosophy*, London: Macmillan.

Stark, W. (1944) *A History of Economics in its Relation to Social Development*, London: Kegan Paul Trench and Trubner.

Stewart, A. and Cohen, J. (1997) *Figments of Reality: The Evolution of the Curious Mind*, Cambridge: Cambridge University Press.

Stewart, F. (1994) 'Are adjustment policies in Africa consistent with long-run development needs?', in W. Van Der Geest (ed.) *Negotiating Structural Adjustment in Africa*, London: Currey-Heinemann.

Stewart, F. and Streeten, P. (1976) 'New strategies for development', *Oxford Economic Papers* Vol. 28.

Stiglitz, J. (1988) 'Economic organization, information and development', in H. Chenery and T. Srinivasan (eds) (1989) *Handbook of Development Economics*, Amsterdam: North Holland.

Stiglitz, J. (1992) 'The design of financial systems for the newly emerging democracies of Eastern Europe', in C. Clague and G. Rausser (eds) *The Emergence of Market Economies in Eastern Europe*, Cambridge: Cambridge University Press.

Stiles, K. (1990) 'IMF conditionality: coercion or compromise', *World Development* 18(7).

Stinchcombe, A. (1978) *Theoretical Methods in Social History*, New York: Academic Press.

Stone, C. (1986) *Power in the Caribbean Basin*, Philadelphia: Institute for the Study of Human Issues.

Stradling, R., Nector, M. and Baines, B. (eds) (1984) *Teaching Controversial Issues*, London: Edward Arnold.

Strange, S. (1988) 'The future of the American Empire', *Journal of International Affairs* 42(1).

Streeten, P., Burki, S., Haq, M. and Stewart, F. (1981) *First Things First: Meeting Basic Human Needs in Developing Countries*, New York: Oxford University Press.

Summers, R. and Heston, A. (1990) *The Penn World Table (Mark V): An Extended Set of International Comparisons 1950-1967*, Washington: National Bureau of Economic Research.

Sweeney, P. (1998a) 'Migrants poisoned and deported', *Observer* 26 April: 1.

Sweeney, P. (1998b) 'Why are we bankrolling this tyrant', *Observer*, 26 April: 18.

Sweezy, P. (1939) 'Demand under conditions of oligopoly', *Journal of Political Economy* Vol. 47.

Tandon, R. and Darcy de Oliveira, M. (eds) (1995) *Citizens: Strengthening Global Civil Society*, New York: Civicus.

Taylor, M. (1976) *Anarchy and Co-operation*, Chichester: Wiley.

Thomas, S. (1992) 'Political economy of privatisation: Poland, Hungary and Czechoslovakia', in C. Clague (ed.) *Journey Towards Market Reform*, Cambridge: Cambridge University Press.

Thompson, E.P. (1968) *The Making of the English Working Class*, Harmondsworth: Penguin.

Thompson, M. (1992) 'The dynamics of cultural theory and their implications for enterprise culture', in S.H. Heap and A. Ross (eds) *Understanding the Enterprise Culture: Themes in the Work of Mary Douglas*, Edinburgh: Edinburgh University Press.

Tietenberg, T. (1992) *Environmental and Natural Resource Economics*, New York: Harper-Collins.

Todaro, M. (1981) *Economic Development in the Third World*, London: Longman.

Togati, T.D. (1998) *Keynes and the Neoclassical Synthesis*, London: Routledge.

Townsend, H. (ed.) (1971) *Price Theory*, Harmondsworth: Penguin.

Toye, J. (1986) 'Political economy and the analysis of Indian development', *Modern Asian Studies* 22(1).

Treece, D. (1992) 'Why the Earth Summit failed', *International Socialism* No. 56.

Trotsky, L. (1973) *Problems of Everyday Life*, New York: Monad.

Turnbull, R.G.H. (ed.) (1992) *Environmental and Health Impact Assessment of Development Projects*, London: Elsevier.

Turner, R.K. (ed.) (1993) *Sustainable Environmental Economics and Management*, London: Belhaven.

Turner, R.K. (1995) 'Environmental economics and management', in T. O'Riordan (ed.) *Environmental Science for Environmental Management*, London: Longman.

UNCTAD (1996) *Globalization and Liberalization: Effects Of International Relations on Poverty*, Geneva: United Nations Conference on Trade and Development.

UNDP (1997) *Human Development Report 1997*, New York: Oxford University Press.

UNDP (1998) *Human Development Report 1998*, New York: Oxford University Press.

UNIN (1986) *Namibia: Perspectives for National Reconstruction and Development*, Lusaka: United Nations Institute for Namibia.

UNITAS (1987) *El Rol de los ONGs en Bolivia*, La Paz: UNITAS.

Vaillancourt, J.G. (1992) 'Marxism and ecology: more Benedictine than Franciscan', *Capitalism Nature Socialism* 3(1).

Valdéz Gutierrez, G. (1996) *Referentes Conflictuales de la Reforma Cubana*, mimeo.

Vallely, P. (1992) *Promised Lands: Stories of Power and Poverty in the Third World*, London: Fount.

van Duijn, J.J. (1983) *The Long Wave in Economic Life*, London: George Allen & Unwin.

Varoufakis, Y. and Young, D. (eds) (1990) *Conflict in Economics*, London: Harvester.

Vogel, S. (1988) 'Marx and alienation from nature', *Social Theory and Practice* 14(3).

Vrooman, J.R. (1970) *René Descartes*, New York: Putman.

Wager, W. (1991) *The Next Three Futures*, London: Greenwood Press.

Wallerstein, I. (1980) *The Capitalist World Economy*, Cambridge: Cambridge University Press.

Wallerstein, I. (1996a) 'The global picture 1945-90', in T.K. Hopkins and I. Wallerstein (eds) *The Age of Transition: The Trajectory of the World System, 1945-2025*, London: Zed Books.

Wallerstein I. (1996b) 'The Global possibilities 1990-2025', T.K. Hopkins and I. Wallerstein (eds) *The Age of Transition: The Trajectory of the World System, 1945-2025*, London: Zed Books.

Walters, R. and Blake, D. (1992) *The Politics of Global Economic Relations*, New York: Prentice Hall.

Walton, J. and Seddon, J.D. (1994) *Free Markets and Food Riots: The Politics of Global Adjustment*, Oxford: Blackwell.

Ward, B. (1979) *The Ideal Worlds of Economics*, New York: Basic Books.

Watson, A.J. (1988) 'The Gaia hypothesis – mechanisms and tests', in P. Bunyard and E. Goldsmith (eds) *Gaia: The Thesis, the Mechanisms and the Implications*, Camelford, Cornwall: Wadebridge Ecological Centre.

WDM (1997a) *Pulling Up The Drawbridge*, London: World Development Movement.

WDM (1997b) *A Dangerous Leap in the Dark: Implications of the Multilateral Agreement on Investment*, London: World Development Movement.

WDM (1998) *WDM Update on the MAI Campaign as at 1st May 1998*, London: World Development Movement.

Weber, M. (1974) *The Protestant Ethic and the Spirit of Capitalism*, London: Unwin.

Weeks, J. (1981) *Capital and Exploitation*, London: Edward Arnold.

Weiss, J. (1983) *Industry in Developing Countries*, London: Croom Helm.

Wells, D. (1996) *Environmental Policy*, New Jersey: Prentice Hall.

Wheelock, J. (1992) 'The household in the total economy', in P. Ekins and M. Max-Neef (eds) *Real-Life Economics*, London: Routledge.

White, L. Jnr (1966) 'The historical roots of our ecological crisis', in I.G. Barbour (ed.) (1973) *Western Man and Environmental Ethics: Attitudes Towards Nature and Technology*, Reading, MA: Addison-Wesley.

Wilber, C. (1979) 'Empirical verification and theory selection: the Keynesian-Monetarist debate', *Journal of Economic Affairs* Vol. 13.

Willetts, P. (1978) *The Non-Aligned Movement: The Origin of the Third World Alliance*, London: Pinter.

Williams, M. (1994) *International Economic Organization in the Third World*, London: Harvester/Wheatsheaf.

Williams, R. (1995) 'Socialism and ecology', *Capitalism Nature Socialism* 6(1).

Wilson, E.O. (1980) *Sociobiology*, Cambridge, MA: Belkap Press.

Wilson, J. (1992) 'Socio-economic justice', in P. Ekins and M. Max-Neef (eds) *Real-Life Economics*, London: Routledge.

Wisner, B. (1988) *Power and Need in Africa*, London: Earthscan.

Wolfe, T. (1997) 'Sorry but your should just died', *Independent on Sunday*, 2 February.

Wood, C. (1995) *Environmental Impact Assessment: A Comparative Review*, London: Longman.

Woodhouse, P. (1992) 'Environmental degradation and sustainability', in T. Allen and A. Thomas (eds) *Poverty and Development in the 1990s*, London: Open University Press.

World Bank (1992) *Governance and Development*, Washington: World Bank.

World Bank (1996) *World Development Report: from plan to market*, New York: Oxford University Press.

World Bank (1997) *Annual Report 1997: East Asia and Pacific*, Washington: World Bank.

World Commission on Environment and Development (1987) *Our Common Future*, Oxford: Oxford University Press.

Worster, D. (1985) *Nature's Economy: A History of Ecological Ideas*, Cambridge: Cambridge University Press.

Wright, E.O. (1975) 'Alternative perspectives in the Marxist theory of accumulation and crisis', *Insurgent Sociologist*, Fall 1975.

Yaron, J., Benjamin, M. and Piprek, G. (1997) *Rural Finance: Issues, Design and Best Practice*, Washington: World Bank.

Yaxley, I. (1996) 'Health care as libertarian mechanism, egalitarian system or social process', *Personal Injury* 3(4), November.

Zeitlin, M. (1970) 'Inside Cuba: workers and revolution', *Ramparts*, March.

Zohar, D. and Marshall, I. (1994) *The Quantum Society: Mind, Physics and a New Social Vision*, London: Flamingo.

Zukar, G. (1980) *The Dancing Wu Li Masters: An Overview of the New Physics*, London: Fontana.

Index